PROBLEMS
IN TRANSFRONTIER
POLLUTION

Record of a Seminar on
Economic and Legal Aspects of
Transfrontier Pollution
held at the OECD in August 1972

ORGANISATION FOR ECONOMIC CO-OPERATION AND DEVELOPMENT

The Organisation for Economic Co-operation and Development (OECD) was set up under a Convention signed in Paris on 14th December, 1960, which provides that the OECD shall promote policies designed :

— *to achieve the highest sustainable economic growth and employment and a rising standard of living in Member countries, while maintaining financial stability, and thus to contribute to the development of the world economy;*
— *to contribute to sound economic expansion in Member as well as non-member countries in the process of economic development;*
— *to contribute to the expansion of world trade on a multilateral, non-discriminatory basis in accordance with international obligations.*

The Members of OECD are Australia, Austria, Belgium, Canada, Denmark, Finland, France, the Federal Republic of Germany, Greece, Iceland, Ireland, Italy, Japan, Luxembourg, the Netherlands, New Zealand, Norway, Portugal, Spain, Sweden, Switzerland, Turkey, the United Kingdom and the United States.

*
* *

FOREWORD

In August 1972, the Environment Directorate organised a Seminar on transfrontier pollution and invited economists, lawyers and engineers from seven countries to present papers on the economic, institutional and legal aspects of transfrontier pollution. This volume contains all of the papers written for this Seminar. These papers were revised as a result of discussions and provide an up-to-date analysis of an area which was briefly examined in "Problems of Environmental Economics".(OECD, Paris 1972).

The multidisciplinary approach to transfrontier pollution appears to be very fruitful and is a factor in the decision to launch further activities at the OECD on administrative, legal and institutional aspects of transfrontier pollution. The papers in this volume have already proved to be very useful reference material for current studies on transfrontier pollution and they will no doubt help in the derivation of practical guidelines for action at the international level aimed at solving transfrontier pollution problems.

The discussion held during the Seminar clearly shows that the transfrontier pollution problem calls for action between the States concerned which bear some form of responsibility for the environmental damage inflicted upon other States. An objective of such action could be the joint management of common or shared environments for the mutual benefit of the parties concerned, recognizing in particular the principle of non-discrimination. Money transfers or transfers in kind between countries would help in achieving a level of transfrontier pollution which would be more acceptable to all parties. Such a level could be taken as the optimal pollution level if the damage costs were known. The implementation of an agreement on transfrontier pollution would be made easier if special institutions were set up and if these institutions were to make use of appropriate instruments for dealing with international environmental economics.

CONTENTS

ECONOMIC ASPECTS OF TRANSNATIONAL POLLUTION

by

Anthony Scott

University of British Columbia, Canada

I. COMMON PROPERTY RESOURCE QUALITY

Most of the economic literature behind the pollution issue is concerned with "externalities", the unintended response of one firm's production (or one person's leisure) to the activity of others. This approach,
drawn from the law of torts and damages, is useful because it integrates
well with the rest of economic analysis. In the best known examples,
concerned with foundries and laundries, the firms are unrestrained
users of the atmosphere, which carries undesired smoke away from the
foundry and alters the fresh air desired for drying wet clothing. The
foundry has private marginal costs and extra social, or external marginal costs of production, that spill over from it to add to the private
costs of the other industry. In other examples, the external effects are
felt within the same industry, by other firms; such as in fishing, petro-
leum, and ground-water drilling. The general approach of this paper is
that it is usually less costly to reform the management of the common-
property resource over which the firm's activity spills, than it is to
intrude into the management of the firms, and their costs. This position
is asserted, rather than proved.

The best intuitive demonstration of the position is to be found in
international affairs. Natural resources which are open to access from
more than one country - on the high seas, by nationals of all countries -
are the subject of today's outcry by environmentalists and conservationists.
Although it would appear, from recent work on oil spills, and on eutrophica-
tion, that it is possible to apply world-wide measures that will prevent
further degradation of the receptor resources, it is equally clear that
it is quite impossible, in the 1970s, to override internal policies of the
countries bordering each shared resource.

What must be done is to concentrate on bringing about adequate
agreement on the management of each shared resource, instead of attempting even to bring about uniform national legislation on all emissions or discharges, regardless of the location or the extent of the pollution they eventually create. The latter sounds just, swift, direct, and
complete. But nationalism and sovereignty, the territoriality of people,
is just as much a modern fact as any other aspect of economy. To seek
environmental quality by attacking the individuality of the nation-state,
in our own life-times, would be inviting defeat. If we wish to clean up
open-access resources, we had best begin with them, not with the policies
and constitutions of the riparian states.

N.B. I am grateful for comments from Ralph d'Arge, Clifford Russell, John Cumberland and
William Baumol, and the research assistance of Shanon Grauer.

Even domestically, however, attention to a resource is probably wiser than attention to the externality which threatens it. A resource is a tangible thing which can be assigned as property by a relatively simple "agency" with rights and powers which have clear motives and clear limits. Thus the agency need not develop policies concerning labour, marketing, finance, or even industrial location. It can go far towards the role of a profit-making (or rent-maximizing) landlord. In the common-property situations mentioned above, it would try to get the largest discounted flow of rentals or royalties, thus taking into account both the present demand and the user cost for the present stock or flow of resource services.

The agency's chief responsibilities are to master the resource, understand its biological and technical potentials and its thresholds, make sure that it is used by those whose utilization is valued most, that a correct balance between present and future use is maintained, and that users' benefits or bids are not reduced by wasteful or redundant harvesting and exploiting facilities.

All this applies directly to resources used for the disposal of wastes. When property rights do not exist, those who use these resources cannot gain by preserving or husbanding them. Instead, they discharge wastes into water or air as long as doing so is more profitable than other ways of combining inputs to produce selected outputs.

This freedom of discharging has three kinds of effects. First, it causes a mis-allocation of all resources towards products, techniques and raw materials that produce wastes and away from those products, techniques and recycling processes that would be chosen if waste discharges were expensive or scarce. Second, both in the long run and in the short run, it reduces the quality of resources enjoyed by those who value the water as an amenity. Third, both in the long run and in the short run, it may actually increase the costs of waste discharges, by choking or exhausting the resource's assimilative capacity.

II.   POLITICAL ACTION AND INTERNATIONAL AGREEMENT

These effects are as important for transnational pollution as they are for a single country. But the "rent-maximization" function now becomes implicit. A central agency, poised between the nations, may find it useful to guess what each nation would pay for cleaner water, but it is unlikely to be able to measure willingness-to-pay directly. Indeed, it is argued in an Appendix to this paper, that no-one, domestic or international, can get this knowledge with sufficient confidence. Hence the rôle of the central administration will likely be that of international secretary, manager, monitor and policeman, rather than that of decision-maker. The member nations may aspire to maximize rent, but all they will be observed to do is pursue certain resource-quality goals.

It follows that, once the types of transnational pollution have been classified, we would be best advised to confine our agency to the modest role of carrying out the pollution and quality notions of an international consensus.

10

Transnational pollution

We may begin directly with transnational pollution. Our procedure is simply to imagine that a common-property "receptor" resource is divided among several nations. Each nation has "dischargers" who dispose of wastes in the resource, and a "public" who benefit from the cleanness of the resource. We note three cases, in increasing order of complexity.

a) All variables in one country. If the resource is not in law or fact divided, but is entirely within one country, the most usual verdict in international law is to say that no trans-boundary situation exists. All the dischargers, all the "public", and all the resource, can apparently be reached by one government.

Unfortunately, the matter is not this simple, for part of the "public", that suffers from a decline in resource quality may be abroad, never directly enjoying the resource. Two routes exist by which foreigners may be injured.

First, people abroad may be offended by pollution outside their own country just as they are now offended on learning that outerspace is becoming littered with jettisoned rocket components, or that seals near Labrador are being cruelly slaughtered. Television broadcasts, magazines, and sermons seem to have made the enjoyment of these resources international, as is shown by the internationally defensive attitudes of the governments responsible for regulating them.

Second, customers abroad may suffer from the higher costs of goods and services exported from polluted regions. This increase, the international "incidence" of domestic pollution, can be positive or negative. (It also provokes the adjustments from customers and rival producers that are the subject of the balance-of-payments studies of d'Arge and others). The incidence may become a burden either if goods made in polluted environments are more costly (because of abatement expense) or have lower quality (e. g. Hollywood cowboy films no longer showing clear California skies), or both. Contrariwise, it may be a benefit, the usual case, if waste disposal or pollution abatement costs are not loaded onto exports but are borne domestically, and then gradually diffused nationally and internationally through changes in the terms of trade.

In all such cases, such burdens as may be felt in adjoining countries will be almost imperceptible to measurement techniques. Consequently, although attention is being given to the trade effects of domestic policies, none goes to the exported burden of these policies, after all market adjustment has taken place. Yet it is quite possible that the residual rise in the cost of Japanese, European or American equipment to users in clean-air LDCs is more worthy of international attention than is the adjustment mechanism by which this burden is minimized. All the output of developed countries may have pollution abatement devices installed, once domestic law calls for them, if costs so dictate. The increased price to foreign buyers is then simply a spillover of pollution costs in the manufacturing country, even when this pollution is entirely domestic.

It is, however, difficult to suggest how much either of these two sources of pollution or abatement costs can be kept at home. Usually

11

they are unrecognized. Their diminution almost certainly would call for foreign pressure on domestic legislation and policy, and would be regarded as interference in what appeared to be a completely internal issue.

If such intervention is feasible, the discussion of the following two cases will apply.

b) Upstream-downstream pollution. The waste dischargers may be in one country, and the "public" in a second. This is a one-way spillover. [1] The enormous literature on externalities, spillovers, nuisance, injunction and compensation applies here.

Within one country, the courts may be used either to establish that one party or group has rights, which can then be marketed or negotiated; or to establish how the resource shall be used or divided. A one-way externality may actually require eventual adjustment by both parties or groups; in welfare economics, mutual adjustment is nearly always called for. Thus each activity or group will benefit less, or cost more, in the joint, optimum occupation of a resource, than if only one of them used it. Law is concerned with which of the two parties, if either, must be compensated for this loss of benefit or increase in cost, while economics is more concerned with the total value of the net production of both groups (including public enjoyment of the resource's amenity). Many important works in economics (e. g. Coase, Buchanan) have been concerned to show that the same final combination of outputs can be achieved, whoever was originally endowed with rights to use the common property's disposal facilities.

With international resources an important difference arises. The gainers are in a different country from the losers. The gross magnitudes cannot, usually, be balanced off against one another. So neither government can be indifferent to the gross magnitudes involved, in its own country. Compensation, in some form or another may be required to make both upstream and downstream parties better off than if no action were taken, and bring home to them the net effects of common action.

Baumol, in his Wicksell paper does suggest the use of the tariff by the losing country to reduce the pollution from a steel mill that also spills into the importing country. But these are rare circumstances. In general, one would expect that the situation will be that the waste dischargers are upstream and the "public" are downstream, and there is no obvious trade or factor flow between the two groups. Then eventual reduction of the waste discharge will only take place if the downstream country offers something else to the upstream country; or if the upstream country, out of prudence or amity, unilaterally decides to make a gift of resource-quality benefits to the downstream public. The compensation, of course, need not be embodied in anything special to the frontier region where the wastes are discharged. It can be a general trade concession, immigration policy, or capital grant; or it may be the offer of a similar pollution policy at another frontier position. Such proxies for direct negotiation or personal compensation certainly render it unlikely that the final adjustment by the two groups will be very similar to that described as "Pareto-optimal" within one country.

1. See Scott: The Economics of the International Transmission of Pollution, Problems of Environmental Economics, OECD, Paris, 1972.

These remarks imply that large numbers of people are involved. In circumstances where those downstream are small in number, they may be able to by-pass their own government by being given direct recourse to the courts of the country where the discharge takes place. Three possible illustrations may be suggested.

   i) International maritime oil-spills conventions appear to envisage that injured parties may sue ships or owners in the owners' own country.

   ii) The Boundary Waters Treaty between Canada and the United States contains language concerning upstream water diversions (not pollution) in which injured downstream parties may apply to the upstream courts for possible redress.

   iii) The new nuisance laws of the State of Michigan make it possible, apparently, for foreigners to use the Michigan courts to enjoin Michigan polluters harming riparians elsewhere.

But these are exceptional circumstances and provisions. Usually much blunter instruments must be used.

   c) <u>Reciprocal pollution</u>.   The waste dischargers may be in more than one country, and the "public" may be also in more than one country. The open-access resource, while a continuous, homogeneous and indivisible medium, may be shared.

This is surely the transnational pollution about which the world should be concerned, for it presents not only intricate economic and social complexities but also the cumulating dangers of world-wide environmental deterioration. This case brings to light several generalizations about the most desirable way for practical economists to approach transnational pollution.

The first of these generalizations is that it would be well for economists concerned with the destruction of an environment to turn their attention to the management of that environment, and away from "externalities". Admittedly, the two approaches, within a single economy, come to the same thing, as was suggested in Part I. The legal tradition (which envisages one party suing another for damages, or the second contracting with the first to refrain from damaging behaviour) however, has been more easily expanded and generalized than an alternative approach which focusses attention on the <u>medium</u> by which the wastes of one firm were transmuted into the damages suffered by another. Hence the various kinds of voluntary agreement which (in the absence of transactions costs) might lead to optimal adjustment of final production between two firms are rarely envisaged as being about a <u>place</u>, or even about a barrier between two "places". Indeed, they are usually not even about new kinds of property right in the "place". Instead they are about actions and practices within the two firms' premises.

This space-less habit of theory cannot easily be adapted to the international problem, however well it works for domestic policy-making. The idea that two persons or firms might bargain until they hit on the right amount of two or more outputs and effluents would not apply well when they are separated in sovereign nations. Their governments would negotiate instead, as principals, and not simply as agents. Each would have

13

to negotiate in such a way as to give due weight to all the interests of all its citizens, not only in a particular international pollution issue, but also in its ramifications for all other present and future "foreign-policy" negotiations on trade, defense, transport or capital movements.

Governments are unlikely to show much willingness to act in this role on behalf of particular citizens. They are also unlikely to have frequent recourse to international arbitration or mediation (on an ad hoc basis) on particular causes, although examples do exist. Much more acceptable is the delegation of a few of the complexities of decision-making for shared common resources to special, permanent, bi- or multi-national agencies. In being separate from their parents, such agencies can come to views, and decisions on their own problems, that are quite distinct from their parent governments' views and policies on other matters. They are perceived to be independent, not part of any-one's current foreign policy. A related advantage is that they can go even further than the parent governments would in enquiring into and indicating the correct quality of some particular natural resource.

For example, they might be able to get some information about costs - both of abatement and of production. Now the international ex-change of useful cost information, when it may lead to adjustments in employment or sales, is rarely accomplished successfully. Although the nations, in defense procurement discussions, in negotiating trade agreements, and in forming cartels have had very good reasons for exchanging such information in the past, they have rarely actually done so. They are prevented by habits of sovereignty and nationalism. At best they have negotiated about more-or-less tangible goals and services, and rarely about the efficient assignment of production between countries. But it is quite possible that, for limited purposes, an international agency limited to a particular resource would have greater success.

Such action would indeed be their main task, but it is not impos-sible that their presence could also encourage side-arrangements be-tween complainants and dischargers. Such side-arrangements would have some of the nature of the adjustments predicted in the Coase-extern-alities models. They might take the form of the beneficiaries paying the dischargers who installed additional abatement equipment, or even the opposite: dischargers paying those who would use the polluted resource. (But the latter settlement would be possible only if the agency was given rather unfamiliar terms of reference, allowing it to act either as pro-moter of water quality or of referee between the two parties.)

In any case, the first generalization is that adversary or diplo-matic proceedings between two governments about the quality of a re-source are inferior to the working of an agency divorced from general foreign policy and free to seek information on damages and costs.

The second generalization, of course, is that international trans-actions costs will be very high. To follow a listing by Mishan and others, the following must be considered in deciding whether to proceed to a final agreement:

i) Initial transactions costs: Identifying the other parties; con-vincing them there is something to talk about; persuading them to agree; persuading non-parties not to enter the resource;

ii) Information necessary for maintaining and revising the initial agreement;

14

iii) Monitoring and policing performance by members.

These are all real costs. Economists should recognize that they may be higher than with domestic resources, and may condemn the world to less precise environmental standards on international resources than on domestic resources. It is tempting to guess that the standards will be lower; but it may be that, once agreement has been initiated, the parties will avoid the costs of later revision and will shoot for harsher standards. There is scope here for the study of a broad international agreement (transcending particular resources) that may reduce the transactions costs of agreement on particular resources.

The third generalization is that international negotiation while forced to seek agreement about arbitrary standards diplomatically negotiated, is not likely to do much worse than domestic agencies setting standards for an internal resource. Once practical economists realize they will never be given the task of comparing costs and benefits in the member countries, in order to strike a cosmopolitan optimum, the object of their economic research becomes very similar to what it always has been. Weights must be given to the gains and losses accruing to their own nationals for various national levels of effluent. Such weighting, presumably derived from distributional ethics and political considerations, is a difficult assignment, and is only recently being attempted within a few countries. What is important is that the economists will not be asked to weight the welfare of other countries, or of entities within it. In international bargaining, each country looks after itself.

The fourth, and final, generalization is that there will often be many countries involved. It is important to remember this. National policies are usually derived from welfare theory which is essentially based on two-party interactions, and many practical economists are accustomed to thinking of curves describing two essentially homogeneous groups: polluters, and the public. But in international negotiations these two parties are of importance only in our case (b), upstream-downstream pollution. In case (c), reciprocal-pollution, the two traditional parties are sub-divided into interest groups, or demanders, in each nation; then married to form as many "national interests" as there are nations.

These four generalizations imply a rough-and-ready procedure for the management of an international resource, that may worry those who seek precision in setting environmental resources. They have little need for concern. Even within one country, standards must be set at levels that are essentially arbitrary. It is not clear that levels set by even less precise methods will be more wrong or less correct.

The justification for this statement is set out at length in the appendix. There it is argued that the demand for pollution abatement, within one country, comes essentially from those who enjoy water quality as a public good, as an amenity. The task of economists advising a government, therefore, is to determine the best scale for a public good, bearing in mind that dischargers will also value the resource's being at least pure enough as to allow them easy disposal of their wastes, and that the costs of prevention or abatement probably can be ascertained. Both the theory and the practice of public finance suggest that economists, so far, simply cannot ascertain the required optimum scale. Hence, the scale of pollution abatement that any team of economists would recommend internationally is unlikely to have emerged from more difficulties of measurement and reconciliation than would be encountered domestically.

This can be expanded by considering for a particular resource, the "demand" and "supply" and their usefulness in deriving the choices of environmental quality.

"Demand" combines that of dischargers for a place to dump their wastes, and that of the public.

Dividing the waste-dischargers up among several countries will not make them demand more waste-disposal facilities than before. Streams and airsheds will still be adequate for them. Even if a market existed, waste-disposers would not be competing among each other and no scarcity price for the resource would emerge. Among them ships, shore establishments, sewage works, would always find enough water to carry their effluents away from their collection systems.

Consequently, in the management of international resources, demand for quality will come predominately from the amenity and health-enjoying "public". The economists' problems of determining the demand price for scarcity of quality will be precisely what they would have been if all the resource were within one country. Basically, the willingness-to-pay for amenity is still diffused among members of a large club. While now scattered among several countries, they are difficult to serve chiefly because what they want is a public good, not because they are internationally dispersed. Any reader of the journal literature on public goods (or Bohm's contribution [1]) will realize that to measure all or part of the aggregate demand for such a public good is a staggering task.

The international "supply" of quality, on the other hand, is little more difficult to determine than domestically. The costs of alternative disposal means (new products, recycling, new techniques, new fuels and materials, abatement works) can still be calculated as within one country. If it is found that the best way of supplying resource quality is by large works (sewage disposal networks, stream abatement, river improvement), the optimal design of these works can be done by a fairly-elementary procedure. (Cf Pearse and Scott 1971. [1])

But data about costs are all that can be obtained. It produces the same information as that for one country, and reduces to one sloping supply curve, disaggregated, if desired, into national and industrial sub-components. Knowledge of this function is helpful, but it is not sufficiant to determine any particular choice of environmental quality. Obviously, the quality to be achieved, measured perhaps in D. O. , (either for a whole resource, or for each part of it for each month of the year) must be made (with a knowledge of these supply prices) by political judgement, based on hearings, questionnaires, elections, editorials, protests and complaints. The diplomats will have the double difficulty of helping their political masters to interpret the demand for quality, and interpreting their masters' doubts and uncertainties. No scheme of rights, charges, or regulations will lead, via trial and error to the "right" level of quality, either in domestic or in international common property resources, so long as there are many parties and so long as one of the demands is for a public good.

1. See, Problems of Environmental Economics, OECD, Paris, 1972.

This is a negative conclusion, but it is helpful. It means that the several international techniques that have been proposed are no more arbitrary in their determination of the final resource-quality objective than is the optimal domestic technique now conceivable. These techniques, to be listed below, all consist of means of rationing the use of the resource for waste disposal, among dischargers.

The whole international procedure consists of the following steps:

1) Initial agreement on need for joint action to supply water quality.

2) Special conference to establish how water-quality standards for each segment of an international resource are to be set, and how to be revised or strengthened, outside diplomatic channels.

3) Implementation of (2) (the decisions reached in the Conference).

4) Near-simultaneous conference to determine how the cost of supplying water quality (or how rights to discharge) will be allocated among the nations.

The first three steps are familiar in international negotiation. Expert conferences, summit meetings, and horse-trading may be necessary. What is of more interest to economists is step (4). The participants here will have to decide on one of five grand "principles" - or perhaps combine them. The aim of the principles is to build up a calculation which will show how much the waste dischargers in each country will share an amount of effluent; and so an amount of D.O. reduction, in the resource. As one alternative, the principles listed in Table I might be applied equally to waste dischargers regardless of their location and nationality (as is done in oil spill conventions). In such procedures the international treaty seems to look through the border to the industries and municipalities themselves. Alternatively, the negotiators may informally calculate what the cut-back in waste discharge from each country would be if a certain principle were followed. Then the remaining amount can be assigned to the country as its waste-disposal "quota", to be passed out among firms and towns as it wishes.

These two routes are polar extremes. The former, as represented by Table I's line (a), tends to "economic efficiency", and would be typified by international agreements to apply identical taxes, subsidies, or regulations to their nationals. (Such uniform policies are the subject of the paper by John Cumberland.[1]) Some readers of an earlier draft of this paper have argued that if it is possible to confer and agree on the joint management of a resource, it should be possible to go farther and agree on the actual steps to be taken by people and firms. No doubt this is sometimes true. But it should be pointed out that most nations sharing an open-access resource will not wish to upset all their internal traditions and arrangements simply to stop pollution of a particular resource. They must also make resource-management policies for other, similar

---

[1]. See, Problems of Environmental Economics, OECD, Paris, 1972.

17

resources; must have concern for the independence and responsibility of their local governments; must give attention to macroeconomic fiscal, monetary and trade policies for income and employment; and must be concerned for the vertical and horizontal distribution of income throughout their entire country. They will probably be very unwilling to change the implied "constitution" of institutions, fiscal instruments and regulatory powers simply to bring about the efficient management of one natural resource, unless they can be shown that there is no other way to achieve joint international action. Each will prefer to set about its assigned resource clean-up task in its own way, one that is least disturbing to the achievement of its domestic policy goals. The countries are not after all, setting up a confederation or a common market. I am driven, therefore, to stress the potential of national quotas which leave each nation free to pursue their own preferences for anti-pollution instruments.

Table I.   SUPPLY OF WATER QUALITY
(allocation of rights of waste disposal among nations)

PRINCIPLE	DESCRIPTION	COMMENT
(a) Economic Efficiency	Marginal cost of alternative waste disposal equalized for dischargers in all nations	Minimizes total cost of achieving chosen quality
(b) Equal economic opportunity	Each unit of population, or GNP, assigned an equal quota of waste disposal	Accepts past build-up, not future economic growth
(c) Equal restriction	Present waste discharges cut back by same amount, or percentage, in all nations	More costly than (b) similar to international effluent standards
(d) Equal behaviour	All dischargers, in all countries, required to use "best available abatement technology"	Requires more consideration than (c), but may be less costly
(e) Equal shares	Each country, regardless of population or GNP, gets a 1/nth share of waste disposal	Rejects past build-up as criterion

Line (e) of the Table, which mentions the principle of equal shares, suggests this quota approach. The quotas may be only implicit. For example, if two countries agree on a set of specific abatement measures that will improve the shared resource's quality, their actions amount to an assignment of the total assimilative capacity between the two countries, and, as negotiations proceed, both will gradually recognize this. One may enquire if it might "substitute" one source of emission, or one kind of treatment, for another, the final effect on the total resource being the same. If they agree, each recognizes that a quota system exists. Or, if they agree on certain specific abatement measures, but add that neither will introduce new effluents in parts of the resource that are so far unpolluted, they have actually divided a fixed total discharge between themselves, and quotas exist.

Consequently, I suggest that economic policy should focus attention on national quotas. The aim of the helpful economist should be to assure the nations involved (i) by reducing their discharges, induce others to do so also, so that (ii) each is made better off than it would be by going it alone, by pressing for the principle or principles in Table I that are most acceptable.

The Consequences of Quotas

Thereafter, each nation can allocate its quota among its waste-dischargers as it wishes, playing favourites, following a list of social priorities, setting charges (following Baumol and Oates, 1971), or selling property rights (following John Dales). Assume that each charges for, or sells, effluent rights, what are the consequences for the location of industry in the countries concerned?

First, it must be recognized that some industries will be unwilling to pay anything for abatement or effluent rights, and will prefer to locate in other parts of the world, adopt entirely new raw materials and processes, or simply go out of business.

The remaining industries will find that their waste-disposal costs, reflecting the new effluent charges or prices, have increased more in some countries than in others, depending on the extent to which each national quota or share is demanded by existing dischargers. If waste-disposal and production techniques are approximately the same at all locations along the shore, industries will actually be forced to redistribute themselves among the riparian nations in search of unused or inexpensive effluent rights. This redistribution would be costly both in re-location expenses and in higher annual resource costs, and consumers and factors of production would have to share the burden of these costs.

Economists will fear that some of these costs are an unnecessary part of the international effort to rise the quality of the natural resource, since they simply arise in moving from one position to another on the same open-access resource - every part of which, we now assume, has the same assimilative capacity.

As such, they would be evidence of "inefficiency" for those who take a one-world view of ideal resource allocation. But they would be regarded as regrettable but unavoidable, like transactions costs, by

19

those who see the drawing of frontiers and the assigning of resource quotas as akin to the assignment of property right in a domestic open-access resource. Both are instruments that delineate initial resource endowments. The migration and re-establishment expenses are simply necessary costs of re-shaping world production in a way that not only allows the member nations to keep the quality of a certain shared resource at a desired level but also recognizes that when this resource's services becomes scarce, it produces a rent, which must be distributed among those nations who have some claim to it. In brief, a new scarcity changes the value of resource endowments, comparative advantages, and the distribution of world income.

There is therefore nothing inherently undesirable in the migration of industry in search of rights to dispose of wastes in a given international resource. The economist, however, may well ask whether such change in locational patterns around a given lake or sea is actually necessary even for the contemplated redistribution. Why should an industry, seeking to rent a nation's quota of discharging rights, actually move to that nation? Could it not just as well pay royalties for these rights while staying in its preferred location, just as it might pay for rights to use a patented production technique, or a management service? If rental income is all that a nation wishes to obtain from its share of the resource, the answer might well be affirmative: the world could save labour and capital if rights were simply rented out.

But such economy of labour and capital - such efficiency - is not necessarily high on national priority lists. Economists should not assume that every riparian prefers income to employment. (For example, in the analogous case of rights to international fisheries, it has been suggested to some nations that they might wish to acquire rights, then sell or rent them to the fleets of other maritime nations. Such suggestions are greeted with hoots of derision.) Most economic policy concerned with nations' management of natural resources seems still to be concerned with creating jobs, a lesson learned by tariff analysts years ago. A government that chooses to become a rentier rather than a developer will not be re-elected.

Nothing essential to the above analysis would be changed if the assimilative capacity of the resource varied from one area to another. This can be seen by considering an example. In a certain lake, the effect on over-all D.O. of 100 tons of effluent dumped into the east end is half that of dumping them into the west end. Other things equal, the riparian national at the east end has greater assimilative capacity, close at hand, than the other nations. This fact may be recognized by other nations and may be influential in affecting that nation's assigned quota for waste disposal; or it may not be recognized, so that the nation is able to dump rather more waste than its assigned quota would suggest. Either way, the nation has a fixed quota. Its price of disposal rights reflects the demand for that quota, and everything else in the paragraphs about location follows: fixed quotas, equal or not, will affect the location of production.

To conclude these brief comments note the following. Costs are all important in making the choices among the five "principles". They differ chiefly:

a)   in the total costs they impose on the nations combined, and

b)   in the costs (of alternative waste measures) imposed on each nation.

20

The benefits to be obtained from the implementation of the principles may be much the same. In any case, benefits cannot be measured, and must be assumed to emerge when <u>political</u> decisions have been taken.

There is hardly any scope for a market decision, with the possible exception of the use of the courts to deal with particular short-run episodes. National laws, enforcing an international convention, with international short-run notification and long-run surveillance, will always be necessary. (For independent evidence that costs, not benefits, are important data, see Russell and Landsberg, 1971.)

Appendix

THE SUPPLY AND DEMAND FOR RESOURCE QUALITY
ON A COMMON PROPERTY

In this appendix, the familiar tools of supply and demand are adapted to indicate the equilibrium amount of purity in a resource, and to argue that the common-property nature of the demand for quality makes it very unlikely that the value of a marginal amount of purity in a given resource can ever be observed, or measured.

Freedom of discharging has three kinds of effects. First, it causes an allocation of all resources towards products, techniques and raw materials that produce wastes and away from those products, techniques and recycling processes that would be chosen if waste discharges were expensive or scarce. These effects are not our business here (although manipulation of them is often regarded as a means of preventing free discharging). By definition they disappear if mismanagement disappears.

Second, free discharging has a pair of effects on the public's enjoyment of the resource. Although the public may (water and air supply and household waste disposal) make use of a resource in a manner which is both technically and economically indistinguishable from that of industry, the aim here is to stress communal enjoyment of the resource as an amenity.

Waste discharge can reduce the scale or value of this public good, with phenomena variously known as disamenities, pollution, and health hazards.

a) "Episodes" or "contingencies"

In the short run, free discharge may temporarily reduce the amenity or health value of a resource. These episodes are becoming very familiar, leading to official closure either of the discharger or of the resource.

b) "Death" or "Destruction of a resource"

In the long run, the cumulation of wastes may destroy the public amenity. They may happen gradually, the resource becoming steadily less attractive; or it may happen sharply, the resource's waste burden suddenly reaching a threshold or saturation point, the resource's environmental, or the public's psychological reaction, being irreversible.

Third, free discharging has a pair of effects which are of chief importance to the dischargers themselves.

c) "Congestion"

In the short-run, free discharge may cause "congestion" of the resource, so that the wastes are not carried away satisfactorily.

d) "Depletion"

Seen from a long-run point of view, today's discharges "depleting" the ultimate capacity of the resource to receive or assimilate the cumulated wastes of a long period.

It is hard to find examples of these two effects. Because congestion occurs only when waste flows are greater than would be tolerated in a public stream from a health or amenity point of view, it is likely to be prevented before it becomes a source of cost to dischargers. Similarly, because a resource is likely to have become destroyed from a health or amenity point of view before its capacity to accept more waste has been used up, depletion of a common-property resource waste-disposal capacity is rarely observed. (A good idea of the two phenomena is provided by a garbage dump, to which dischargers have unrestricted access. On certain periods the dump may be "congested", in that no more visitors can use it in a day. However, it is not thereby used up; it may last many years more. Each discharge however "depletes" its ultimate capacity, bringing closer the day when it can accept no more.)

The small table below shows these two pairs of effects on the two classes of resource user. The important distinction is between the two columns. The public's enjoyment of a resource, and its dislike of pollution, is not based on a competition for enjoyment between members of the public. Up to a very large population, a river or fresh air can be enjoyed without congestion or depletion. Private, waste-disposal utilization on the other hand, is competitive among dischargers.

Table II.   EFFECT OF FREE WASTE DISPOSAL

	PUBLIC ENJOYMENT (Public good)	WASTE DISCHARGERS (Common-property)
Short-run	(a) Pollution episodes and contingencies	(c) "Congestion" of waste-absorptive capacity
Long-run	(b) Cumulative and irreversible destruction of amenity	(d) "Depletion" of waste-absorptive capacity

Industrial, versus public demand for quality.   Casual observation of
private and group behaviour in search of improved resource quality
suggests that the public's loss of amenity, not the dischargers' need
for waste disposal in the river, is behind most protest, and intervention. The following familiar diagram may make the point precise. The
curves lie above the quality (D. O.) axis, along which is measured fraction
of D. O. saturation from zero (very heavy waste loading, odours, and
ugliness) to 100% (no pollution). The vertical axis measures the money
value of a one-percent change in quality.

The lowest, solid-line demand curve shows the increment in benefit from a gain in quality, as seen by a waste discharger. Because such
a firm does not suffer from, say, the anaerobic degradation of the river,
the incremental benefits fall to zero at very low levels of quality.

Figure 1

COSTS AND BENEFITS (SUPPLY AND DEMAND)
OF WATER QUALITY AT A SPECIFIC LOCATION

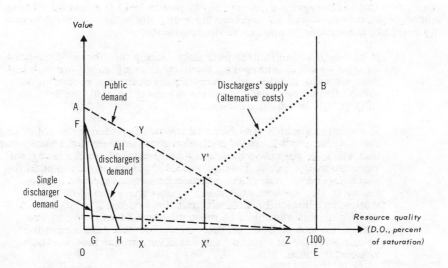

The upper, solid-line demand curve shows the horizontal summation of such demand, or benefit curves, for a large number of discharging firms.

The dashed demand curve is the vertical sum of the demands by
members of the public for increased quality of water. Each and all suffer
as soon as D. O. falls much below 100.

Hence, the use of a stream for waste discharge (shown by moving
to X from Z) withdraws no stream benefits from those who are already
discharging. Even if rights to this amount of the resource were for sale,
they would not be scarce, and waste dischargers would not bid for them.

25

Only the public suffers. Its demand curve tells us that it is, under pressure, willing to pay up to the triangle XYZ to restore the D.O. to 100%. In the absence of jurisdiction, and the organisation to intervene, it protests, as long as protest costs less than XYZ. (For a discussion of these two demands for water quality, see also Kneese, Ayres and d'Arge, Chapter 3.)

The Supply of quality.   Of course, there is a cost to increasing quality. New products, processes or materials must be introduced. The dotted line is the incremental cost curve of raising the D.O. of the stream from X upwards to 100%. This is the summation of the alternative-cost curves of all the discharging firms, added horizontally from right to left, in terms of their respective reductions of the resource's D.O. The area under this dotted curve measures their total benefit (cost-saving) from discharging their wastes into the resource.

A final decision about level of D.O. can now be determined on a variety of sets of assumptions. All suggest that, subject to distributional considerations, X' is the optimal state of the river. For example, if the river were private property and the owner were able to sell its waste-discharge and amenity services, it would pay him to exclude wastes as long as the public were willing to offer more for a further increase in D.O. than the dischargers were willing to pay to hold it constant. The landlord can be envisaged as maximizing rent, the total amount depending upon his monopolistic powers to discriminate.

a)   If he could discriminate perfectly among members of the public and waste dischargers, then, by making an all-or-nothing deal with the public, and charging dischargers according to their alternative costs of waste disposal he could collect up to OAY'BE in rental revenue.

b)   If he could exclude members of the public, then he could charge them a flat rental, and the dischargers another flat rental, per unit of D.O. He could maximize his rental when the marginal revenue in the public market equalled the marginal cost in the dischargers' group. This solution however is ruled out because D.O., or non-pollution, is a public good. Not only is exclusion physically difficult, but the product, purity, is indivisible. All members of the public would be enjoying the entire amount of D.O. in the river after the discharges had been cut back, not simply the smaller amount that was being offered to them.

c)   The same difficulty rules out charging a single price higher than Y'. If the two groups demanded divisible products, such a price would maximize rent if it were as close as possible to that level where the marginal demand price and the marginal supply price were equal to zero, and to each other. (This three-way equality can only be approximated, at the best of times.) But again, the fact that the demanders consume a public good makes this solution impossible.

d)   Thus the maximum rental, in the absence of discrimination, must be the price that clears the market. This is Y', giving a rent-maximizing D.O. level of X'. Note that, as usual, this is the same solution that would be reached by a perfectly-discriminating monopolist.

It is also the level of D.O. that would be reached if a Dales-type rights market existed. If it were possible to sell units of D.O., the dischargers and "public" would compete until a price of Y' cleared the rights market, at X' D.O.

Furthermore, if there were complete integration so that a sole owner controlled the resource, the discharging firms, and the "public", the total of waste-disposal-cost savings and consumer benefits would be maximized if the level X' were chosen.

On various assumptions, X' is also the social or Pareto optimum level of waste disposal. The sum of marginal rates of substitution between private income and the public good would be equal to the marginal social cost of waste abatement. Consequently, any other arrangement would reduce the output of goods and amenity below what is socially possible, or make one consumer of these two products worse off.

However, there is no point in trying to set up a market to determine the optimum degree of pollution, as long as one of the demands is for a public good. The literature of public finance, stressing non-excludability, zero marginal costs and indivisibility all point to the necessity of a political process, not a market. The most that can be hoped for is that a public authority, a large "club" in James Buchanan's terminology, will enter the market which will be competitive because there are also a large number of rival, would-be waste dischargers. (Then the public would be like a hunting club entering the competitive farm land market in order to reserve nesting grounds for game birds. The club, like the public, does have free riders. But the whole hunting group is sufficiently small to be, at most, just one competitor in the land market.) Thus, whether or not its outcome would be optimal, the market route to determining the degree of each resource's pollution might not function at all.

The analysis suggests therefore that the political task is to reach a degree of pollution that is the same as, or analogous to X'. Should it be sought? The literature has placed five warnings before us, in addition to that about public goods in the paragraph above; each suggests that X' may not be an optimum.

a)   The theory of second best: because demand prices may not be equal to marginal costs elsewhere in the economy, the movement of the political process toward X' may actually reduce total welfare. This warning should be taken very seriously. It is quite possible, for example, that reducing pollution on a particular river to X' may increase pollution on other, unregulated or uncontrolled resources. Other effects of policy, such as the increase in monopoly are more indirect but in sum may be even more undesirable.

b)   Ignorance about and variability of alternatives: because we cannot usually directly connect waste disposal by one firm with a whole resource's change in D.O., the supply curve XB may be impossible to plot. No-one may possess the required information. In any case its shape will vary between positions on a river, or in the atmosphere. Only if all dischargers are located close to each other on a single lake or sea can a single supply curve be drawn leading to a single optimum X' for the entire resource. Another reason for this is that the supply curve must be drawn relative to its own demand curve, even though the benefits of D.O. may be experienced some miles away, downstream.

27

c)   Ignorance about willingness to pay: the demand for D.O. is a public good; and it is notoriously difficult to discover the demand for such a good. (Cf. Bohm, Swedish Journal and OECD papers.) Further, such curves depend on, and shift with experience of alternative resource conditions. They are not therefore independent of the amount of D.O. in existence.

d)   Irrelevance of willingness to pay: because the willingness to pay must be focussed on the demand of those who have some knowledge of and taste for the amenities of the resource, it may neglect the demand for unpolluted resources of those who are not directly involved. Their more remote demand for D.O. should also be considered: preservation values, option demand and the isolation paradox are all relevant here. (Weisbrod, Krutilla, Sen). Perhaps purity of the resource, or the welfare of flora and fauna, should be treated as a merit good. (Musgrave.)

e)   Maldistribution of income: because social income distribution enters not only into the willingness of the "public" to demand amenity, but also into the final derived demand for the goods which produce wastes, and their by-products, the apparent solution at X' may be far from a social optimum. It appears impossible, though, to judge whether, if income were more evenly distributed, the result would be for X' to be higher or lower.

Further, even if all incomes are equal, because public action usually does not involve compensation to those who lose, but proceeds to balance gains against losses, it may quite arbitrarily redistribute wealth without regard to rights or merit.

CONCLUSION

We may combine a comment on these difficulties with a conclusion to this part of the paper.

It is becoming traditional, in economics, to admit that these are daunting difficulties about welfare economics, and policy making; then to proceed anyway. This must be the procedure here. What these difficulties do is:

a)   remind us not to make major or minor reforms without considering whether they really are second-best to total reform of the economy; and

b)   remind us to examine carefully the distributional consequences of any policy.

Furthermore we should remember that most pollution is the change in nature of a social amenity, a public good. Hence, even if it is positive it cannot be expected to prevail against private waste-disposal use of common-property, even if the common property has a market for rights to use it (à la Dales); because it is a public good, the demand for it can be manifested only through social or public - political - action.

Finally, because the total demand for a public good is never manifested or revealed in a market, the strength or intensity of public demand is not manifested, except politically or in protest. It is sometimes said of pollution that there are two relevant shadow prices, and that both must be met. These are said to be the price offered by those who value greater purity for its amenity, and the price offered by those who value greater purity for its capacity to dispose of wastes. This generalization is indeed true, but unhelpful, for neither price ever appears.

The latter "price" is, as the diagram showed, usually zero (waste disposers do not usually need a clean resource for their discharges). The former price, although it has conceptual existence, is simply unobservable.

More helpful is the following statement. When, by political action the purity of the resource has been moved to some position like X', there will be observable a marginal cost of further private waste disposal by alternative means. Allowing for the five or six welfare-economics problems on the previous page, it is desirable that in an optimum this observable marginal cost (such as Y') be equal to the marginal social value of greater resource purity. When introspection and questionnaires suggest that a change to greater purification is worth this cost (of abatement) then the optimum has not yet been reached. When no improvement is deemed to be possible, the observed cost of abatement becomes a measure also of the social demand for D. O.

The amount of pollution, or purity, in a resource must be determined by political action. Except in those rare cases where the benefit is entirely private (as for commercial fishing) or is institutionalized into a relatively small demand for a divisible resource (as with Ducks Unlimited in the land market) no market procedure can lead to a position like X'. In the context of this conclusion, we may briefly examine the instruments for resource-quality improvement, so often discussed in text-book literature. Variously classified, these amount to: (a) charges for reduction in resource quality (effluents) (b) transferable rights to discharge. Both these can be seen simply as means of rationing the reduction of D. O. from E to X' among those dischargers who, if the common property were free, would prefer to discharge enough to reduce D. O. to X. Some of these means are much more efficient than others, and they have been very carefully examined. Notice however, that even under ideal circumstances neither of them works automatically, nor helps to locate X'.

It is true that varying the charge for pollution (a) will help to determine the position and slope of the supply curve. But it does so by placing the burden on the firm (and its customers). This incidence of the cost makes it more difficult for the legislature - or the public (Bohm) - to compare benefit with cost. The theory of public finance suggests that, as for the scale of a public good, the best way to locate X' would be for the public, through proposing varying levels of taxation on itself, to raise that revenue that would be used to subsidize waste dischargers to use alternative methods. But the literature is all too full of warnings about how subsidies encourage waste and evasion.

In any case, we must agree with those who have concluded that X' must first be laid down by fiat. This having been done, either of the two fiscal rationing devices above may be used, to cut down discharges. In addition, it is possible to bring into use

    c)   regulations about quality of emissions

d) regulation of fuels, equipment and recycling techniques, or

e) taxes or subsidies on discharge-reducing materials.

These are less general and may be more expensive, for they give the dischargers and the public very little chance to optimize. Thus they are of even less help than (a) and (b) in discovering the position and the slope of the supply curve. But as they are similar to those rationing devices already in use in sanitation, safety, public health, quarantine and land law, they are acceptable, recognizable and direct, even if expensive in their working.

Finally we should recall that it is traditionally the courts and litigation that settle disputed questions about interference to the enjoyment of an asset. This recourse is still available, it should be listed:

f) legal procedures for injunction and compensation.

These have been well described and discussed by Burrows for British practice (OEP, 22, 1 March 1970, 39-56), and doubtless some parallels exist in other systems. In general, the law and the courts seem to accomplish two things. First, they establish rights to use a resource or to be free from interference with one's enjoyment of it. These rights can become the basis for subsequent bargaining whereby the party who has rights redistributes them between himself and others; this bargaining can be two-sided, as a continuation or complement to a legal action for rights, or multi-sided, as in the markets sketched above. Second, especially in administrative law, a court or board may be assigned to determine how much right each party has to use or enjoy. The first function would be performed if a court decided that the public, say had a right for the D.O. level to be set at E, and that the public and the dischargers were free to bargain their way to points like X or X'. Performing this function, the court merely makes valid and enforceable the transactions that may take place in a market. The second function would tend to settle the whole matter by placing the parties at a point like X', without the right to negotiate a redistribution of the rights to some other level of D.O.

The first function is not really important to our brief discussion here. In setting the stage for bilateral (or internalized) bargaining procedures, it is an essential foundation for what has already been noted. Further, if dischargers vie with each other for the waste-disposal services of a resource, the first function of the courts is to form a basis for there to be some rationing of discharging rights.

The second function, however is more important. It may take the place of the political process in setting the amount of pollution in the resource. It is not simply a rationing device, like the instruments numbered (a) to (e), and the first function of the courts, above.

REFERENCES

Wm. J. Baumol and W. E. Oates, "The Use of Standards and Prices to Protect the Environment", Swedish Journal of Economics, 73 (March 1971), pp. 42-54.

R. H. Coase, "The Problem of Social Cost", Journal of Law and Economics, (October 1960), pp. 1-44.

P. H. Pearse and Anthony Scott, Joint Selection and Optimization of Projects by Two Jurisdictions (unpublished).

Marc J. Roberts, "Organizing Water Pollution Control: The Scope and Structure of River Basin Authorities", Public Policy, 19 (Winter 1971), pp. 79-141.

C. S. Russell and Hans H. Landsberg, "International Environmental Problems - A Taxonomy", Science, 172 (25 June 1971), pp. 1307-1314.

THE ECONOMICS OF UNIDIRECTIONAL TRANSFRONTIER POLLUTION [1]

by

Gilberto Muraro

University of Padua, Italy

1. The present paper, originally presented to the OECD in May 1972, has been modified in order to incorporate the paper written for the OECD Seminar on Transfrontier Pollution: "The Cost-Sharing Approach to Upstream-Downstream Transfrontier Pollution" (AEU/ENV/72.12). The author thanks Prof. S. Ch. Kolm, Dr. W. Rungaldier, Prof. A. D. Scott, Dr. H. Smets, and the members of the Central Analysis and Evaluation Unit (OECD Environment Directorate) for their helpful remarks and suggestions. The author remains, of course, fully responsible for anything that may still be wrong.

INTRODUCTION [1]

For the purpose of this paper, transfrontier pollution (TFP hereafter) is defined as pollution originating in one country and affecting another country's environment, through natural media.

This definition does not encompass all the cases of exported pollution (since pollution can also be exported by movements of goods and services), nor, a fortiori, all the "international environmental problems" involved (since an international problem can arise from pollution affecting, through natural media, the environment of an international zone like the seas, the oceans or the high atmosphere; and it can arise also from a case of pollution physically confined within national boundaries, either because it affects the existing pattern of international trade, or even because there is an option demand for the amenities of the polluted country in other countries. [2] It simply identifies a special class of such problems, on the implicit assumption that it is possible and convenient to analyse them in isolation.

Transfrontier pollution may be unidirectional or reciprocal (including in the latter case pollution of a lake or of a river shared by two or more countries). This paper will deal with the former case, i.e. with the economic problem of "one-way externality". [3]

The analysis shows that one-way TFP does not call for a new economic theory in so far as the optimum targets and instruments are concerned. The particular complications arise on the political side, being linked with the sovereignty of the parties and the lack of an accepted statement of environmental rights in the bargaining process. Indeed,

1. Throughout this paper the following abbreviations will be used:

TFP = transfrontier pollution
CSA = cost-sharing approach
PPP = polluter pays principle
VPP = victim pays principle.

As regards the PPP, please note that it can have different meanings: the polluter may be required to pay only the abatement cost necessary to meet some fixed emission standard (as in the OECD version), or he may be required to pay in addition either the residual damage costs (as prescribed by international law), or a "tax" related to the amount of residual pollution discharged (as in the market-simulating solution advocated in economic literature). The principle is also consistent with lumpsum transfers to the polluter. When it is relevant to the analysis, this paper specifies the particular meaning intended.

2. For the importance of option demand and preservation values for environmental optimum policy, see I.V. Krutilla, "Conservation Reconsidered", American Economic Review, September 1967.

3. On the same topic see H. Smets "Alternative Economic Policies of Unidirectional Transfrontier Pollution", this book p. 75. As regards "reciprocal externality", i.e. the pollution of a common property, see A.D. Scott, "The Economics of the International Transmission of Pollution" in Problems of Environmental Economics, OECD, Paris, 1972; and "Economic Aspects of Transnational Pollution, this book p. 7.

the "polluter pays principle" (PPP hereafter) has not yet been implemented with regard to TFP; at the same time, since that principle is
advocated by international law and there is a growing moral pressure
in the international community for its overall implementation, the polluted country is being rational in refusing to accept the "victim pays
principle" (VPP hereafter).[1] The search for a solution which is both
efficient and politically feasible has shown that no help can be obtained
from the principles of "equality" so often mentioned in discussing environmental economics, such as "equal environmental standards",
"equal emission standards", and "equal marginal cost of pollution
abatement". If one accepts existing international law as a datum, the
only available course of action consists, paradoxically, in backing the
PPP in the hope that the final outcome will be, not the implementation
of that principle, but the adoption of a cost-sharing approach in which
the polluted country bears the residual damage and a part of the abatement cost.

It is argued, however, that it would be preferable to make a change
in the law so as to enable the liability of the polluter to be made a function of the level of his economy.

In the rather long Appendix to this paper an attempt is made to
give analytical support to the views expressed in the text.

I.   THE NATURE OF THE PROBLEM

As has been said in the introduction, one-way TFP means the
pollution which one country (B) receives from another (A) through air
or water flow.[2] The repercussions can of course involve many countries, imposing external diseconomies on one another seriatim.

The intensity of TFP will be either the same as that of the original
pollution in the upstream country, or different from it, depending on the
relation between the two national environments. Generally speaking, we
can say that the quantity of waste discharged in A which crosses the
frontier, after having produced in A a certain level of pollution q^A, will
produce in B a pollution of intensity $m\,q^A$ where m represents the relation between the assimilative or dilutive capacities of A's environment and
B's environment. In the case of water pollution by non-degradable substances, we can directly consider m as the ratio between the flow rates
of the river in A and in B.

The TFP may be either separable or inseparable from B's internal
pollution. The situation where it is inseparable will be considered here
in its simplest but commonest form, namely its additive form. When
TFP is non-additive B will have two separate damage functions, but
when TFP is additive, the damages in B must be calculated from the

1.   It must be underlined that this statement with regard to international law plays a crucial
role in our analysis. It has been taken from the paper by Ch. Bo Bramsen, "Transnational Pollution and
International Law", this book p. 257.
2.   For the sake of simplicity we will speak of countries, even when actually only a limited
area in each country is concerned with TFP.

total quantity ($q^T = q^B + mq^A$) without the possibility of distinguishing between the two components. As regards the cost of pollution abatement, one must in any case preserve a distinction between the two elements, because, while B can always reduce the national component of its pollution by controlling its emissions, it may well happen that elimination is technically or economically impossible in the case of TFP. This seems always to be the case with air pollution, and it is also the case with water pollution in all those instances in which the large size of a river precludes the idea of carrying out pollution abatement "inside" it (e. g. by re-oxygenation, or treating the whole river, or varying the quantity).

In practice this means that the downstream country may be prevented from achieving the desired environmental quantity, no matter how many resources it is ready to devote to the task, if the downstream country does not co-operate by cutting down its own emissions. It also suggests that the approach of combatting TFP by means of large-scale works organised on a joint basis can find application only in a comparatively few, though not irrelevant, cases. Generally the only way to deal with the problem is by a stricter control in the upstream country, and that is the approach considered in this paper.

As for the other technical and economic aspects of the problem, it is enough to recall that TFP can exhibit all the well-known variations of national pollution. Thus we can have TFP with degradable and non-degradable pollutants, reversible and irreversible deteriorations, independent or synergistic forms of pollution, and so on.

II. THE OPTIMUM SOLUTION AND ITS THEORETICAL IMPLEMENTATION

Let us assume for a moment that the costs of pollution abatement and the damage from the residual pollution (having regard also to their values in the future) are known in all the countries concerned. Then a definite solution may be found which is the optimum from an international point of view in the sense that it maximizes the welfare of all the countries involved.

Such a "global optimum" requires that in each country the pollution shall be kept down to the level at which the marginal cost of pollution abatement is equal to the sum of the marginal damage costs incurred inside the same country and in the downstream countries.

Considering the case of two countries, such a global optimum solution inevitably requires a reduction in the quantity of pollution to below the level which suits the upstream country from the point of view of its national interest alone. As regards the downstream country, no change is called for by the national optimum solution when the TFP is of the separable type. Even with an additive TFP the level of pollution abatement in B which was the optimum from a domestic point of view with the existing TFP will not be increased in the global optimum situation, if we assume, as it seems realistic to do, that the damage increases proportionately or more than proportionately with the quantity of pollution. In the first case - corresponding to a constant marginal damage -

the level of abatement will remain unchanged; in the second case - corresponding to an increasing marginal damage - the level will decrease. We shall readily agree with the latter conclusion when we realise that the downstream country will have had to cut its own emissions drastically (so incurring high costs) in order to compensate for the flow of TFP and that the international approach leads precisely to alleviating the cost to the downstream country by reducing the TFP.

In any case, then, the total quantity of pollution generated in a global optimum situation will be less than the total quantity generated under independent decisions taken inside national boundaries.

Passing from the target to the instruments, it will be seen from the Appendix that all the instruments available for dealing with national pollution problems are, as a matter of principle, also available for solving the TFP problem optimally. It is just a question of bringing the variables to their new level of equilibrium. That means, with reference to the polluting country, imposing stricter standards on its emissions, if the administrative system is adopted, and putting a higher unit tax on the polluting discharges or making higher unit payments to the polluters for the potential pollution they prevent, if the economic systems are adopted. An original combination of those economic systems can also be used to take account of the international interests involved in the stricter regulation of pollution; it consists in a unit tax on the residual pollution fixed at the level necessary to reach the optimum solution for A (the polluting country), and a unit subsidy on the potential pollution prevented, fixed at the level necessary to induce a move from the national to the global optimum solution. Finally it is theoretically possible to make use of a "Swiss Corporation" as suggested by Scott,[1] i. e. a private agency which the countries involved commission to manage their resources, allowing it to look for maximum rent. Starting from the point of maximum potential pollution in the upstream part of the environment, the agency would sell a certain amount of cleanness to the downstream country and the residual pollution to the upstream country. Under certain assumptions of competitive behaviour, the equilibrium point would correspond to the global optimum situation.

As a matter of principle, the difficulties in implementing the various approaches (as described in A. 3 of the Appendix) are the same in the national case as in the international one.

III. POLITICAL ASPECTS OF TFP AND RELATED ECONOMIC PROBLEMS

In conclusion, one-way TFP does not call for a new economic theory in so far as the optimum targets and instruments are concerned. The particular complications arise on the political side, being connected with the sovereignty of the parties concerned.

The first complication concerns the evaluation of the damage done by the residual pollution, which is tantamount to the benefits from pol-

---

1. See A.D. Scott: "The Economics of the International Transmission of Pollution", in Problems of Environmental Economics, OECD, Paris, 1972.

lution abatement. It must be stressed that the benefits do not only concern the distributional issues, but also contribute to determining the most efficient solution. Now it is well known that the evaluation of those benefits almost always calls for a political judgment, the reason being that in the most important instances the quality of the environment, considered from the consumer's point of view, represents a public good in the strict sense (not only is the consumption of it available to all, but it is also impossible to exclude anyone), so that the collective preferences emerge only in the political process. We cannot master the problem by calling for a wider use of the techniques of analysis for measuring the willingness to pay for the quality of the environment. This is not only because of the huge technical problems, but above all because of the conceptual limitations of the approach in a field where widespread ignorance of the effects of pollution and important considerations concerning the distribution of wealth weaken the basis of individualistic welfare economics. In effect, in some cases it is proper to qualify the quality of the environment, even when preserving the features of a public good, as a merit good, i.e. a good which the government supplies in a larger quantity than that which is desired by the community concerned. This also happens in a national framework, where the target for the quality of the environment is generally set in an arbitrary way and is haphazardly changed through time, depending on the cost of pollution abatement and the political pressures for a higher standard.

An obvious additional complication in the international case is connected with the lack of a central authority that could evaluate the benefits. These become, therefore, implicitly or explicitly, a subject of bargaining.

The second complication concerns the greater importance of the distributional issues in an international framework and especially in connection with TFP. The various approaches to dealing with pollution mentioned above - direct controls, levying charges on polluters, and making payments to polluters - are equally good from an efficiency point of view (in so far as their effects on incomes are not too serious), but they imply different distributions of the burden among the groups involved in protecting the environment. Thus even in a national framework we can imagine the interplay of political pressures aimed at influencing the choice of the method of controlling pollution, but the possible existence of some general law on rights to the environment, some automatic compensation that is likely to be given inside the national system,[1] and the existence of a central authority are factors which make for a relatively smoother debate in the national case than in the international one, where they are missing. So we come to the heart of the problem, which is the bargaining process between sovereign States with regard to TFP.

The starting point in the analysis is the recognition that no principle for fixing the initial rights of the two parties is mutually accepted. It is true that international law advocates the PPP and that there is a growing moral pressure in the international community for applying that principle in international environmental disputes also, but so far the PPP has put up a poor performance in dealing with TFP. All that can be realistically assumed is that international law and moral pressure

---

1. For instance, a payment system which puts the burden primarily on the public budget and is likely to imply the sacrifice of some alternative public expenditure. But the sacrifice suffered by the community at large may be offset, at least in a transitory period, by the avoidance of an increase in industrial prices and of undesired sectoral crises.

may induce the upstream country to forego any rent for selling its pollution abatement activity and only to ask to be paid the net cost (that is the cost of pollution abatement minus the related benefit to the same country). Given certain assumptions, bargaining on that basis would lead to the global optimum situation. The difficulty is that the downstream country will not want at all to start bargaining on that basis, as it will consider that the fair starting point should correspond to zero pollution. Looking at the real world, we can find more than one international environmental dispute in which the parties do not move from their original positions towards an agreed solution.

In analysing the question one must take care not to be misled by the usual reasoning of static welfare economics. In such a static framework the stubbornness of the victim seems to violate the postulate of rationality; one can have all the moral rights one likes, but if one is not able to have them recognized, it is rational that one should pay in order to make the best of a bad situation. Thus, putting the question in a static framework, one should conclude - "à la Candide" - that we are always living in the best of all possible worlds, since the damaged part could have bribed the damaging part, and the fact that it did not do so means that the cost of pollution abatement in the polluting country is higher than the correlated gain in the polluted country, so that the existing pollution corresponds to an optimum situation.[1] It may well be that in some instances the problem takes precisely this form, so that only a distributional issue remains open (whether or not the polluter should compensate the victim), without any question of allocation. However, the empirical studies already available clearly show that in many instances the reduction of TFP promises a net gain from the international viewpoint.

Accordingly, the fact that the victim does not bribe the polluter must be interpreted in a different way. Such behaviour must be placed in a dynamic context, taking into account the fact that there is, as already stated, a growing moral pressure in the international community for the acceptance of the PPP in international as well as purely national disputes. It follows that the stubbornness of the victim in complaining and refusing a compromise may well pay off in terms of future recognition of his rights.

It is likely that in addition non-economic considerations, such as considerations prompted by national pride, will play a part in one-way TFP disputes, but the important point is that the behaviour can be rationally explained.[2,3]

1. This statement requires the additional assumption that the income effects of the pollution are not relevant, otherwise the starting point, i.e. the right or the prohibition to pollute, would influence the optimum solution. On this subject see E.J. Mishan, "Pangloss on Pollution", The Swedish Journal of Economics, n.1, 1971.

2. The situation is much better in cases of reciprocal pollution which offer the possibility of joint benefit from solving the problem, but in cases of one-way externality the downstream country only sees the unfairness of the arrangement whereby it is damaged without damaging the other party, while the upstream country only sees the net cost of additional pollution abatement. Note also that, with reference to water pollution, the upstream country may argue that the downstream position involves a negative rent in respect of water quality which, however, is compensated by a positive rent in respect of water quantity and associated activities (as, for instance, the benefits accruing to countries which possess harbours at the mouth of an international navigable river).

3. In formal economic analysis, non-economic considerations can easily be included under the heading of "psychological" externalities of the consumption type; but this does not really contribute to an understanding of the issue.

To clarify the issue, let us assume for a moment that the TFP originating in country A can be at two levels only, namely the present level q_0, and the level q_1 (with $0 \leq q_1 < q_0$); q_1 does not involve damage to the downstream country B, but of course it calls for the implementation of an appropriate pollution abatement policy. In some cases country B will be rational in refusing to bribe the polluter, as for example:

i) when B is sure that the PPP will be implemented in international disputes by a certain date, with respect to which the present value of the damage is less than the present value of the abatement cost;[1]

ii) when B, confronted with a fixed abatement cost and a fixed damage cost, puts a subjective probability on the early acceptance of the PPP such that the expected value of the damage is less than the value of the abatement cost;[2] and

iii) when, apart from any guesswork as to the future implementation of the PPP, B arrives at a certain subjective probability (p) that A will be induced to eliminate TFP because of some bargaining strategy (connected with trade, transport, defence, finance or any other item) which B will adopt, and the expected saving in pollution abatement costs is higher for B than the cost of that strategy.[3,4]

1. Analytically, the problem may present different features:
   i) fixed cost of pollution abatement policy = C; annual damage = D; then B calculates the time x, for which
   $$C - \sum_{t=1}^{x} D(1+i)^{-t} = 0$$
   The date x is then the break-even point of the "waiting policy"; if B is sure that the PPP will be implemented no later than x, then it is better to wait;
   ii) same as at (i) above, but with an annual cost C instead of a fixed cost. It is assumed that, once B has agreed to pay the abatement cost, it will have to pay it for ever; the formula then becomes
   $$C/i - \sum_{t=1}^{x} D(1+i)^{-t} = 0$$
   where C/i is the present value of the constant flow C.

2. In this case acceptance of the PPP is probable, but not certain. As for the time, in order to simplify the issue one must assume that it is irrelevant, e.g. a situation in which in the short term the TFP could produce a definite damage D, which can be avoided only by the rapid implementation of a pollution abatement policy costing C. Then B puts a subjective probability on the rapid acceptance of the PPP (before the damage is caused) and decides to wait (assuming no particular "risk-aversion" or "risk-love" on the part of B) if
   $$C - pD > 0$$
   This example reflects the case already considered by H. Smets "Contraintes dynamiques dans le problème de la pollution transfrontière", OCDE, 11 July 1972 (Annex II of "Alternative Economic Policies of Unidirectional Transfrontier Pollution", this book p. 75).

3. If E represents the cost of the strategy for B, we have: $pC - E > 0$
   Once again the formula implies risk-neutrality in B's behaviour.

4. Scott and Baumol have already pointed out how bargaining about the international environment may call for the most varied strategies of an economic and political nature. Baumol, for instance, in his Wicksellian Lecture, pointed out that the downstream country B can raise a tariff against imports of the products of A's polluting activity in order to induce A to cut down his pollution, and that "if B is a major importer of the good in question, this approach may be quite effective". See A.D. Scott, "Economic Aspects of Transnational Pollution", this book p. 7, and W.J. Baumol, "On International Problems of the Environment" (in preparation).

41

Cases (i) and (ii) assume that the possible implementation of the PPP is an event completely independent of the behaviour of the downstream country. Such a hypothesis may be valid in some instances, but when the countries involved have a significant voice in international affairs they may play an active role, for instance in the international organisations which support or oppose the implementation of the PPP. Since this active role would imply a cost, the third case may be typical of most present-day situations.

In all three cases the situation is by definition a temporary one which should eventually come to an end, either with the fulfilment of the downstream country's forecast, or with its recognition that its forecast and strategy were wrong and its consequent acceptance of the "victim pays principle". However, the waiting period may be quite lengthy, so that an arrangement designed to shorten it would offer prospects of an improvement in international welfare.

In seeking such an arrangement we shall now analyse two propositions:

i) the cost-sharing approach;

ii) the applicability of some guiding principles of pollution control to the problem of unidirectional TFP, as advocated by writers on the subject or by the international organisations.

IV.   THE COST-SHARING APPROACH TO TFP

In view of the political difficulties in implementing either the "polluter pays principle" or the "victim pays principle", one is driven to look for a compromise between these two polar principles. One of the obvious compromises consists in the cost-sharing approach (CSA hereafter), whereby the downstream country would bear the cost of any possible residual damage and a part of the pollution abatement cost.[1]

Let us, then, analyse, in an heuristic way, the mechanism of the CSA.

4.1. The CSA mechanism when there is a definite alternative level of pollution

Given our interpretation of the present situation in probabilistic terms, the working of the CSA must likewise be viewed in this frame-

1. A different approach - already analysed by H. Smets in "Alternative Economic Policies of Unidirectional Transfrontier Pollution", this book p. 75 consists in sharing T = C+D, where C represents the abatement costs and D the residual damage. In that approach, assuming fair behaviour on both sides in assessing C and D, any cost-sharing basis will be compatible with efficiency, so that the choice will be governed only by distributional considerations; this is so because in any event both countries will find it convenient to minimize the total amount to be shared. The approach implies, of course, mutual agreement on the damage function, which is not necessary in the CSA studied in the present paper.

work. Let us continue, for the time being, to assume that there can only be one other definite level of pollution as an alternative to the present one (this hypothesis is less restrictive than it seems; it may well be that the damage function has a critical value such that, if something has to be done, it must consist in reducing pollution to that value).

Considering the downstream country B it is easy to conclude that in all the three cases mentioned above there will be a share of the cost (let us call it bC) which is less than the expected cost of the policy of waiting (pD or E), so that B will be ready to participate immediately in a pollution abatement policy on that basis (always assuming risk-neutrality on B's side).

Let us now consider the upstream country A. In the first two cases (PPP likely to be implemented in the future) if the alternatives were to do everything or nothing, A would invariably prefer to do nothing, i. e. to wait until it was really compelled to implement the pollution abatement policy at its own expense. However, there is a certain maximum share of the cost for A as well - let us call it aC - which is less than the expected value of the total future cost. In the third case, when B threatens as a revenge to take a step that would inflict certain damage on A, A would find it convenient either to bear that damage or, if less costly, to implement the pollution abatement policy and bear the cost C or, finally, if less costly, to apply some countermeasure which would put a stop to B's strategy and would impose a cost G on A (with G < C); but here, too, there is a share aC which represents the minimum outlay for A.

In conclusion, when there is only one single relevant level of pollution as an alternative to the existing one, there are always certain maximum shares, a and b, of the pollution abatement cost that are acceptable to A and B respectively. The obvious consequences are:

i) when a + b = 1, there is only one single solution;

ii) when a + b > 1, various solutions are possible and the bargaining will be limited to the distribution of the net gain from the CSA;

iii) when a + b < 1, the dispute cannot be settled by means of the CSA (though the CSA reduces the gap between the two claims). [1]

1. Let us consider briefly some technical problems of the CSA in the two favourable cases (i) and (ii). In a stationary economy, once we assume that the present value of the cost of pollution abatement has been measured (or agreed), there is no particular problem with regard to the CSA. B would pay to A the agreed share, after which A could implement the pollution abatement policy as it wished, by means of direct regulation, or taxes or subsidies, or a combination of taxes and subsidies. In a dynamic setting with changes in the extent and composition of the polluting activities and abatement costs, the CSA would call for continuous recalculation of the burden to be imposed on each country. This calculation would be easier when only a few large polluters were involved and/or when the pollution abatement policy was implemented by means of large-scale operations; but it should not be too difficult even when many polluters are involved, in which case variations in the regulating instrument (standard or tax or subsidy) needed to keep the pollution down to the agreed level would be a great help in calculating the new burden to be shared. However, one can well imagine that the relations between the countries might be regulated once and for all by a lump-sum transfer covering the risk of future variations in the present value of the cost to be shared. This would make it more difficult to reach agreement on the expected present value of the costs; on the other hand, it would avoid future disputes, so that it might well be preferred by both countries. In this connection, a significant example is the lump-sum settlement agreed upon by the United States and Canada for the development of the Columbia River.

Needless to say, formal support for the CSA by international organisations could greatly alter the subjective probability as to the implementation of the PPP and, more generally, it could bring about a change in attitudes to the problem.

4.2. The CSA mechanism when there is a variable level of pollution

Let us now come back to the usual assumption that pollution is a variable and that both the damage and the abatement costs are continuous, strictly monotonic and twice differentiable functions of the upstream pollution.

4.2.1. As a first step it is better to disregard the element of uncertainty, i.e. to disregard any guesswork about the future implementation of the PPP and any threat from the victim. In that framework, assuming that the CSA is accepted, for each basis of cost-sharing which is proposed each country will offer to reduce TFP down to the point where the internal marginal damage equals the partial marginal cost which that country must bear. In other words, each country has a continuous offer line linking the proposed quantity of final pollution to the basis of sharing. As shown in A.5 of the Appendix, the two offer lines meet at a point corresponding to the global optimum solution. This point would be reached by bargaining under conditions of behaviour of the competitive type. Within the most appropriate model of bilateral monopoly another point might be reached, different from the previous one but still the Pareto-optimum. [1]

From what has been said it will be clear that, if a definite basis for cost-sharing were imposed a priori on the two countries, it would lead to different proposals as to the quantity of final pollution (note, however, that the proposed quantities would be intermediate between the initial amount of the TFP and zero TFP), except in the following two cases:

i) when the cost-sharing basis was accompanied by the right, given to a country X, to choose the final quantity of pollution in the upstream country A (where X may be A). Since country X would choose the quantity for which its marginal cost ($a^X c$) was equal to its marginal damage (d^X), that quantity would coincide with the "global optimum quantity" only when the country's share of the cost was equal to the quota of the total marginal damage from TFP suffered by that country. Otherwise the quantity of residual pollution chosen would be larger or smaller than the global optimum. In the former case it could also happen that the excess of the cost of pollution abatement over the benefit from it at the margin would involve a situation worse than the initial one (i.e. with full TFP) from the point of view of international efficiency;

ii) when the proposed quantities coincide. As is shown in A.5 of the Appendix, if one assumes linear marginal functions, this would happen only when the shares of the cost and the marginal damage borne by the countries were in the right proportion.

This second case is the more interesting, since the agreed solution would also be in the optimum one. It is easy to see that this result

---

1. See T. Scitovski, Welfare and Competition, Ch. XIX, G. Allen and Unwin, London, 1957.

is directly in line with the economic theory of public goods, whereby the optimum quantity of a public good (in this case pollution abatement) can be obtained by letting each party pay a price for it equal to his marginal benefit.[1]

This may suggest a way of dealing with the problem which may seem both politically feasible and economically efficient at the same time, consisting in: (a) adopting a cost-sharing approach based on the damages avoided; (b) building up an international technical board with the function of assessing the damages and costs and of proposing an efficient solution; (c) implementing such a solution.

However, this course has two weaknesses. The first is due to the fact that damages cannot be assessed only on a technical basis, as was said in the preceding paragraph, and that once a voice is given to governments in assessing their damages and therefore their demand for the public good, the well-known tendency to understatement will appear.[2] The second and really decisive weakness is due to the fact that a cost-sharing approach based on the damage done involves a political choice at the beginning and therefore does not avoid the extra-economic confrontation which is the main feature of the TFP problem. Indeed, that kind of CSA is tantamount to acceptance of the "victim pays principle" by B. There is fair play on A's part which allows B to pay only the net cost, i.e. the cost of pollution abatement minus the benefit to A. It is exactly the kind of agreement which in the preceding paragraph we considered acceptable to A but not to B.

4.2.2. What we are looking for is a CSA which is more favourable to the victim. Here we must once again introduce either a threat from the downstream country or the probability of compulsory implementation of the PPP in the future.

Without recourse to a formal illustration, it is sufficient to say that in such a framework A's offer line will be modified in a sense favourable to B and that for each proposed share, A will be ready to accept a greater reduction of pollution than before.[3] As far as B is concerned, the starting point is its rejection of the victim pays principle when there is a choice between accepting that principle and pursuing a waiting policy; but it would accept the idea of cost-sharing, should this be proposed. That means that B has an offer line which intercepts A's offer line, so that an agreement is eventually possible (unless non-economic factors, such as national pride mentioned earlier, become relevant).

1. A further well-known qualification is necessary, namely that people should behave in accordance with the model of pure competition; otherwise, if people take into account the marginal loss connected with a higher demand because of the increased price, production would be lower than the optimum. Note also that the price = marginal benefit rule would lead to financial equilibrium by providing the public good only if the marginal cost was constant; otherwise it would lead to a net surplus (with higher marginal cost) or to a deficit (with lower marginal cost), so requiring additional lump-sum transfers if financial equilibrium had to be preserved.

2. In many cases this weakness may not be an important one, both because the small number of the countries involved is likely to keep this kind of 'free-rider' problem within bounds and because the technical board has a good chance of being able to act as the arbiter in such questions.

3. Let us consider the case in which the pollution is a variable, but the marginal damage function for the upstream country A has a kinked curve, being zero when $q < q_i$ and having a positive and increasing value when $q \geqq q_i$. If one assumes that in the initial situation the TFP is at level q_i corresponding to the internal optimum for A, no cost-sharing arrangement involving a further reduction in TFP will be acceptable to A, unless it is afraid of a future implementation of the PPP or of some retaliatory measure by B. But the CSA would become acceptable to A once it had reason to fear these things. (See A.5 of the Appendix.)

Before drawing any policy conclusions from the CSA, let us consider the second approach, namely the application to TFP of some principle of pollution control advocated in the international framework.

V.   THE APPLICABILITY TO TFP
OF SOME INTERNATIONAL PRINCIPLES OF POLLUTION CONTROL

Some principles are under discussion in international meetings as well as in international organisations with regard to the problem of the impact produced on international trade by policies for preventing pollution. Each one of those principles aims at giving effect to a particular concept of equity and/or of efficiency. Let us now see whether any of them may be recommended for regulating TFP efficiently.

i) Equal marginal cost of pollution abatement for all discharges. If we assume, as it is realistic to do in many instances, that the various techniques of pollution abatement are available everywhere at the same real cost, this principle will be the most efficient one for dealing with the pollution of a common property resource. It is by far less efficient if applied on an international basis in relation to national pollution programmes, when we allow for different damage functions in the various nations because of different environmental endowments and/or different tastes. In our case the optimum solution requires that the marginal cost in the upstream country A should be equal to the sum of the marginal damage in A and of that part of the damage in B that is due to the TFP, while in B it should be equal only to the internal marginal damage: $c^A = d^A + md^B$; $c^B = d^B$. The principle can therefore be efficient only if $d^A + md^B = d^B$, which requires the assimilative capacity of the environment in B to be $1/m$ times greater than in A. When the two regions have the same assimilative capacity, which may happen very often, the principle would certainly be inefficient, unless the marginal damage of pollution in A were equal to zero. It may have some political appeal, however, and it involves a relatively simple system of control when economic methods of regulation are adopted (equal unit tax or equal unit subsidy).

ii) Equal environmental standards. (In an internationally shared resource these are of course an intrinsic feature of the situation and not a guiding principle). As an international rule for national environmental policies this principle is only efficient in particular cases (countries with the same social and economic structure, the same preferences, the same original environmental capacity; or countries with different cost and damage functions that lead, through an internal compensation process, to the same solution). As a general rule it has therefore been criticized in theory,[1] and in practice it is strongly opposed by the developing countries. When applied to TFP it may also prove to be efficient only in particular cases, a necessary condition being that the assimilative capacity of the environment must be higher in the downstream

---

1.   See J.H. Cumberland, "The Role of Uniform Standards in International Environmental Management", and A. Majocchi, "The Impact of Environmental Measures on International Trade: Some Policy Issues" in Problems of Environmental Economics, OECD, Paris, 1972.

country than in the upstream country.[1] Note that in reality the principle is concerned with the target and not with the instrument, so that it minimizes interference in national affairs, leaving each State free to manage its environmental problems as it likes. That may increase its political acceptability, but also makes it more difficult to check compliance with an international convention, since it lacks any reference to operational rules.

iii) Equal emission standards. It is well known that this administrative system does not minimize the total cost of environmental protection (unless we assume equal cost functions for pollution abatement for all discharges). Nevertheless, it has some important applications in national environmental policies, because of its supposed equity and of its lower cost for purposes of decision-making and monitoring. For the same reasons it may have a strong appeal in reciprocal TFP disputes. As a general rule for international application it is open, to a greater degree, to all the theoretical and political objections already considered in connection with principle 2. In the problem of one-way TFP the necessary condition for the rule to be compatible with an efficient solution is, once again, a higher assimilative capacity of the environment in the downstream country or a lower level of maximum potential emission in the upstream country. In the case of two countries with the same assimilative capacity and the same potential emission (which is likely to imply equal populations, equal incomes and equal productive structures) the principle becomes equated with principle 2, whose inefficiency has already been shown. [2]

In conclusion, there is no principle which is invariably efficient or, at least, invariably superior to the others. Note in addition that all the principles considered above do no more than establish rules of the game, i. e. constraints on the final solution, and that they do not establish initial rights; thus in themselves they are not sufficient to lead with certainty to an agreed solution.

VI. CONCLUSIONS OF THE ANALYSIS
AND THEIR POLICY IMPLICATIONS FOR THE OECD

Let us sum up the conclusions of the analysis and see what role in TFP disputes one can suggest that the international organisations, and particularly the OECD, should play.

6.1. Each of the two polar principles - "polluter pays" and "victim pays" - would be a sufficient and efficient principle for solving the TFP

1. In A.2 of the Appendix it is shown that in the global optimum situation - assuming the same quadratic expression for the cost function and the same linear expression for the damage function in A and in B, and assuming furthermore the same assimilative capacity of the environment in the two countries - the total pollution in B should be more than twice the pollution in A.

2. Another principle is considered by Scott as potentially applicable to the reciprocal TFP problem. It consists in sharing among the interested countries an agreed total quantity of waste discharged into the common resource. But it is evident that such a rule is irrelevant to one-way TFP, where the discharges of the upstream country affect the welfare of the downstream country, but not vice versa.

problem. The PPP is not yet being implemented in international environmental disputes, but, since it is advocated by international law and there is a growing moral pressure for its implementation, the polluted country is rational in refusing to accept the opposite principle. Thus, from the point of view of international welfare, the problem of TFP calls for an arrangement that could lead rapidly to an acceptable and efficient solution.

6.2. The confrontation between sovereign States involved in the TFP problem may be subject to some constraints stemming from the rules of thumb of the international organisations to which all the countries concerned may belong. For instance, it is difficult to imagine a tariff being used as an anti-TFP instrument inside the framework of the Common Market, or the menace of a political quarrel inside NATO. Although there remains wide scope for tough bargaining, such constraints would seem to improve the situation. Thus it would seem that a positive role may be played by the international organisations, notwithstanding the bilateral relationship involved in one-way TFP.

The most obvious role consists in offering one's services as match-maker and arbiter and in helping to start the bargaining and keep it cool. If these efforts are to succeed, the countries involved must recognize the moral authority and expertise of the international organisation, and for this purpose the regional organisations should be more effective than the global ones. The right course seems to be to choose the smallest of the organisations which includes all the countries concerned with TFP.

This proposal, however, does not take us very far, and a more ambitious course would be to lay down some rules enjoying wide acceptability. But do we have any rules to propose?

6.3. None of the principles of "equity" examined above, namely the the principles of "equal environmental standards", "equal emission standards" and "equal marginal cost of pollution abatement" invariably passes the tests both of economic efficiency and of political feasibility.

As for the PPP hitherto advocated by the OECD and implying that the polluter pays the cost of abatement but not of the residual damage, it cannot automatically be used as a guiding principle for TFP, since it calls for a central authority to fix the standard to be met by the polluter, which is exactly what is missing in TFP. What can be done with that principle is to back it as a constraint in the bargaining. What would happen, however, is that the polluter would respect the PPP to the extent of effecting some reduction in TFP, while the victim would have to pay for any further reduction. In other words, we should realize that the real outcome would probably be a kind of cost-sharing.

As for the CSA studied in this paper, it has been shown that it seems to suit both parties as a means of shortening the waiting period, when due account is taken of the growing pressure for the implementation of the PPP and of the possible countermeasures which might be taken by the polluted country. When there is only one level of TFP alternative to the existing one, there may be circumstances in which agreement is not possible, but they look like being exceptional cases. When the level of TFP is a variable, agreement is always possible. Note, however, that bargaining is necessary, both with regard to the basis of sharing and as regards the quantity of final pollution (similar to price and quantity bargaining in a bilateral monopoly).

6.4. If these, then, are the conclusions of the analysis, what action can one suggest that the OECD should take (in addition to the role of peacemaker already mentioned)?

The possible courses of action open to the OECD would seem to be the following:

i) To back the provisions of international law which make the polluter fully liable in TFP disputes. This would imply following the present trend and allowing individual countries to agree on a CSA, given the probabilistic considerations set out in the foregoing analysis. A variant of this action, with the same kind of result but probably less effective, would be to back the PPP as interpreted by the OECD and to maintain that the liability of the polluter extends only to the abatement costs.

ii) To give strong support to the CSA in international environmental disputes, without specifying a definite basis of sharing. Perhaps this would raise the question whether it was appropriate for the OECD to back a principle so different from the PPP accepted in international law. But the main point is that such action would aggravate the position of the victim; indeed the expected loss to the polluter and the expected gain to the victim from the "waiting policy" would then be respectively decreased and increased. The polluter, while accepting the CSA, would endeavour to make it meaningless by putting the main burden on the victim. Thus either the victim would actually accept the victim-pays principle (which, as already stated, would be consistent with the CSA when the polluter also suffered damage), or he would have to resort to some more effective strategy of intimidation. We think that neither of these results would be regarded as positive by the OECD, which would probably prefer to take the side of the victim and would surely wish to avoid the use of threats that could weaken co-operation among Member countries.

iii) To back the CSA with a specific sharing rule. The rule could be based on fixed shares, or on shares related to some geographical or economic parameter (e.g. the per capita income of the regions involved in TFP). Apart from the political objections mentioned under (ii) above it is clear that, when the shares are defined a priori, there is still room for disagreement on the final pollution level, unless the fixed shares should by chance coincide with those acceptable to the two countries in the light of the specific data on the problem.

In conclusion, of the three courses of action just outlined we would suggest the first one. That is because we believe that in any case, if an agreement were reached, it would be based on some cost-sharing rule. However, the best way to favour that outcome is to give formal support, not to the cost-sharing approach, but to the polluter-pays principle.

Our conclusions are given empirical support by the recently announced settlement of the dispute regarding the pollution of the Rhine, based on a cost sharing arrangement between France and Germany on one side and the Netherlands on the other side, all of which countries belong to the OECD, which advocates the PPP.

ADDENDUM

A.   International law, noting the full liability of the polluter, exercises
certain moral and political pressures that are crucial both in interpreting
the present situation and in arriving at the policy implications for the
OECD. It is not out of place therefore, to consider whether international
law is exercising the right pressure at the right level, or whether it
might not be appropriate to suggest some other rule.

In our opinion, there is a principle which is both more effective
(in terms of likelihood of being implemented) and more equitable than
the present PPP; a principle which seems to be related to the CSA - and
which in practice may well lead to a CSA - but is actually aimed at stating
the initial rights in TFP bargaining.

B.   This principle can be summarized by the formula "don't do to others
what you don't do to yourself", and it could be stated as follows:

i)   in one-way TFP, the polluter must behave in such a way that
the pollution carried over to the downstream country does not
exceed (unless agreement is reached with the downstream
country) the pollution that the upstream country allows in its
internal regions which, from the standpoint of socio-economic
structure, are similar to the region polluted in the other
country;[1]

ii)   the countries involved in a TFP dispute must set up a commission on which they would be equally represented and which would
have the task of defining what the principle outlined in (i) above
implied in practice in the case in question. In the event of disagreement, the countries would agree to have recourse to, and
to accept the judgment of, international experts appointed by
mutual agreement.

It seems to us that this principle, while compatible with efficiency,
is also acceptable from the political view-point when due consideration
is given to the wide economic differences in the world today. It is sufficient to recall the main feature of the Stockholm Conference - the confrontation between the developed and the developing countries - to realize
that such a principle is much more acceptable in practice than the PPP,

1.   The principle is the same as that advocated by S. Chr. Kolm as a "non-discrimination" principle. In discussing implementation (point (ii)), we put more emphasis on the action of the States concerned
than on private or local action, as did Mr. Kolm. (See S. Ch. Kolm, "Guidelines for the Guidelines in the
Question of Physical International Externalities", this book p. 249.)

because it does not place the entire burden on the polluter when economic conditions are worse in his country.

In addition, the countries involved must start with a preparatory commission responsible for drawing up the terms of reference (pollution in similar internal regions), which is a task that does not involve tough political confrontation. In this way the bargaining process which is to follow may be made easier, because an institution will have already been created and have worked successfully.

Appendix

ANALYTICAL NOTES ON TRANSFRONTIER POLLUTION (TFP)

A.1. Outline of the problem

A.1.1. Simplifying assumptions and general remarks

The TFP problem will be studied on the following assumptions:

i) that the assimilative and dilutive capacity of the environment in each country is constant (no seasonal variations are considered);

ii) that only one type of pollutant is considered, a pollutant that is not assimilated, but only diluted by the environment, because of its chemical nature or because of the irrelevance of self-purification in the basin considered. The degree of pollution then becomes a linear function of the amount of the waste discharged;

iii) that the downstream country cannot decrease the pollution coming from upstream. This assumption is certainly valid for air pollution, but it seems realistic also for most cases of water pollution (e.g. where there is no economic possibility of controlling pollution "inside" the river by varying its flow, re-oxygenation, or treating the whole river);

iv) that the marginal utility of money for the consumers concerned is constant, so that the marginal cost and damage functions do not change, whatever be the initial distribution of the rights.[1]

The following symbols will be used:

V = assimilative capacity of the environment

W = waste discharged

q = degree of pollution due to national waste discharges in the country considered

q_o = maximum potential pollution (corresponding to the private optimum discharge for the polluters when discharge is free)

---

1. In section A.4, devoted to the bargaining process, the hypothesis of a decreasing marginal utility of money will also be considered.

53

C and c = respectively, total and marginal cost of pollution abatement

D and d = respectively, total and marginal damage caused by existing pollution.

Capital letters as superscript will indicate the countries (A, B, C), or the total pollution (T, S).

Apart from the hypotheses set out above, the problem may present various aspects, depending on:

i) the number of countries involved, i.e. whether two or more (countries A, B, C, etc., along the river or course of the air-flow);

ii) the relation between the assimilative capacities of the national environments, i.e. whether they are equal or different. In the first case $V^A = V^B = V^C = V$. In the second case $V^A = mV^B$; $V^B = nV^C$, and so on. The most frequent case, and the only one considered here, concerns an increasing assimilative capacity of the environment as one goes downstream (for instance, because of the increasing flow of an international river), so that $0 < m < 1$; $0 < n < 1$;

iii) the cost function for pollution abatement and the damage function for residual pollution: we assume here the irrelevance of the fixed values and non decreasing marginal values for both functions. [1]

iv) the relationship between TFP and the internal pollution in the downstream country. If these two components are not separable, there will be only one single damage function. In the particular example of non-separability considered here, namely additivity, it will be $D = D$ (total pollution), where total pollution is equal to the sum of the TFP and the internal pollution. If they are separable, two distinct damage functions will exist: $D_1 = D_1$ (TFP), and $D_2 = D_2$ (internal pollution).

Note that in any case the link between the national environmental sub-systems is provided by the damage function in the downstream countries, as will be seen from the following description.

A.1.2.   Case of two countries

A.1.2.1.   Taking two countries with the same assimilative environmental capacity, and with additive pollution and constant or increasing marginal cost and damage, the descriptive system is as follows:

---

1. However, the combination of constant marginal cost and constant marginal damage will be excluded, since it would lead to an unrealistic extreme solution (zero or total pollution abatement). As regards the damage from pollution, note that it seems to show increasing or constant marginal values until a critical point is reached (for instance, a degree of pollution that does not admit of any further recreational activity), after which it shows a decreasing marginal value (See G. Brown Jr. and Brian Mar, Dynamic Economic Efficiency of Water Quality Standards or Charges, Water Resources Research, Dec. 1968). It is realistic to assume, however, that in the range relevant to the optimum solution the marginal damage will be increasing or constant.

$$q^A = \frac{W^A}{V} \qquad\qquad q^B = \frac{W^B}{V}$$

$$c^A = c^A(q_o{}^A - q^A) \quad \text{with } c^A = \frac{dC^A}{d(q_o{}^A - q^A)} > 0 \;; \quad \frac{d^2C^A}{d(q_o{}^A - q^A)^2} > 0$$

$$c^B = c^B(q_o{}^B - q^B) \quad \text{with } c^B = \frac{dC^B}{d(q_o{}^B - q^B)} > 0 \;; \quad \frac{d^2C^B}{d(q_o{}^B - q^B)^2} > 0$$

$$D^A = D^A(q^A) \qquad \text{with } d^A = \frac{dD^A}{dq^A} > 0 \;; \quad \frac{d^2D^A}{d\,q^{A2}} \geqq 0$$

$$D^B = D^B(q^T)\Big|_{q^T = q^A + q^B}$$
$$\text{with } d^B = \frac{d\,D^B}{d\,q^T} > 0 \;; \quad \frac{d^2D^B}{d\,q^{T2}} \geqq 0$$

A.1.2.2. Assuming a higher assimilative capacity for the environment in B than in A, other things remaining equal, we have:

$$V^A/V^B = m \;; \quad 0 < m < 1$$

$$q^T = (W^A + W^B)/V^B = mW^A/V^A + W^B/V^B = m\,q^A + q^B$$

Thus the damage function for B changes as follows:

$$D^B = D^B(q^T)\Big|_{q^T = mq^A + q^B}$$

A.1.2.3. When TFP and internal pollution are separable, other things remaining equal, the damage function for B changes as follows:

$$D^B = D_1^B (q^A) + D_2^B (q^B)$$

A.1.3. Case of more than two countries

The extension of the problem to the case in which more than two countries are involved is straightforward, it being sufficient to add new relationships of the type already given. For instance, taking three countries, we have to add a new damage function which will assume the following form in the various cases dealt with above:

A.1.3.1. (same assimilative capacity of the environment)

$$D^C = D^C(q^S)\Big|_{q^S = q^A + q^B + q^C}$$

55

A.1.3.2.   (increasing assimilative capacity of the environment)

$$D^C = D^C(q^S) \Big|_{q^S = mnq^A + nq^B + q^C}$$

where

$$m = V^A/V^B, \quad n = V^B/V^C, \text{ so that } mn = V^A/V^C$$

A.1.3.3.   (non-additive pollution; one type of TFP originating in A and carried into B and C, and a second type originating in B and carried into C)

$$D^C = D_1^C (q^A) + D_2^C (q^B) + D_3^C (q^C)$$

A.2.   The optimum solution

A.2.1.   Individual optimum solution

For country A the individual optimum solution is represented by q_i^A, where $c^A(q_i^A) = d^A(q_i^A)$.

For country B, the individual optimum solution, once A has reached its equilibrium point, is represented by q_i^B, where $c^B(q_i^B) = d^B(q_i^T)$;

where $q_i^T = q_i^A + q_i^B$ if the assimilative capacity of the environment is the same in A as in B ($V^A = V^B$);

and $q_i^T = mq_i^A + q_i^B$ if $V^A = mV^B$.

If TFP and national pollution are not additive, there will be two distinct damage functions. With regard to TFP, country B cannot ex hypothesi help itself, so that it will remain burdened with $D_1^B (q_i^A)$; its own internal pollution will then constitute a quite distinct problem and B will find what can be called the individual autonomous optimum solution in the quantity q_{ia}^B, where $c^B(q_{ia}^B) = d_2^B (q_{ia}^B)$.

Needless to say, the latter quantity would also be the equilibrium point for B with additive types of pollution, in the absence of TFP.

Analogous solutions, duly adjusted, are valid for other downstream countries.

A.2.2.   The global optimum solution

As distinct from the above-mentioned individual optimum solutions, one can imagine a global optimum solution, i.e. a solution which maximizes international welfare and which the countries involved would adopt, were they unified.

A. 2. 2. 1.   In the two-country case, it would consist in minimizing the function (given equal assimilative environmental capacities):

$$\underset{(q^A, q^B)}{\text{Min. } Z} = C^A + C^B + D^A + D^B$$

Solving the system formed by the first-order conditions for minimization, we find the global optimum quantities q_g^A and q_g^B : [1]

$$\begin{cases} \dfrac{\partial Z}{\partial q^A} = -c^A + d^A + d^B = 0 \\[2em] \dfrac{\partial Z}{\partial q^B} = -c^B + d^B = 0 \end{cases} \qquad \begin{cases} c^A = d^A + d^B \\[2em] c^B = d^B \end{cases}$$

With different assimilative environmental capacities in A and in B, the global optimum solution requires that

$$\begin{cases} c^A = d^A + m d^B \\[1em] c^B = d^B \end{cases}$$

The expected result is that the marginal damage caused to B by the pollution created in A will be reduced by the coefficient m. Ceteris paribus, this means that in the optimum situation A will be allowed to pollute more, the smaller the value of m.

Finally, with non-additive types of pollution, the function of the Z type depends only on q^A, so that the optimum is at q_g^A, for which $c^A = d^A + d_1^B$ (or $c^A = d^A + m d_1^B$, if $V^A = m V^B$), where d_1^B is the marginal damage in B caused by TFP. Of course B should also reduce its internal pollution to the optimum quantity q_{ia}^B, where $c^B = d_2^B$.

A. 2. 2. 2.   The conditions for a global optimum solution can easily be extended to the case where more than two countries are involved. Taking, for instance, three countries (with the same assimilative environmental capacities), the problem becomes:

$$\underset{(q^A, q^B, q^C)}{\text{Min. } Z} = C^A(q^A) + C^B(q^B) + C^C(q^C) + D^A(q^A) + D^B(q^T = q^A + q^B) + D^C(q^S = q^A + q^B + q^C)$$

Taking the first partial derivatives as equal to zero, we obtain

$$\begin{cases} c^A = d^A + d^B + d^C \\ c^B = d^B + d^C \\ c^C = d^C \end{cases}$$

---

1.   Note that the second-order conditions for minimization are surely satisfied, given our assumptions for the cost and damage functions.

57

When the assimilative capacity of the environment increases as one goes downstream, as indicated at A.1.3.2. above, the conditions for optimality become:

$$\begin{cases} c^A = d^A + md^B + mnd^C \\ c^B = d^B + nd^C \\ c^C = d^C \end{cases}$$

Finally, with non-additive TFP and constant V, the optimum requires:

$$\begin{cases} c^A = d^A + d^B_1 + d^C_1 \\ c^B = d^B_2 + d^C_2 \\ c^C = d^C_3 \end{cases}$$

In conclusion, given n countries involved seriatim in one-way externality, the condition for a global optimum is to keep pollution in each country down to the level at which the marginal cost of pollution abatement is equal to the sum of the marginal damages inflicted on the same country and on the downstream countries (with corrections to the coefficients of the type m, n and mn applied to the marginal damage from TFP when the assimilative capacity of the environment increases as one goes downstream).

A.2.3.   Case of two countries: a diagrammatic exposition

Let us go back to the case of two countries (with equal environmental endowments and give a diagrammatic exposition of the problem. The optimum solutions and their interrelations are shown in fig 1.

A.2.3.1.   The case of non-additive TFP is shown in fig. 1(a), (b) and (c). In this case a country's internal pollution constitutes by definition a completely distinct problem which finds its optimum solution in the quantity q^A_{ia}, where $c^B = d^B_2$ (fig. 1c). The amount of TFP must then be reduced by A to the level q^A_g if a global optimum is to be reached (fig. 1b).

A.2.3.2.   Coming now to the most interesting case of additive TFP, it is desirable to distinguish between the more general case of increasing marginal damage both in A and in B - as shown in fig. 1(d), (e) and (f) - and the particular case of constant marginal damage in B - as shown in fig. 1(g) and (h). The marginal cost is assumed to be increasing. [1]

The results can be summarized as follows:

_____

1.   The case of a constant marginal cost of pollution abatement is not shown, since it gives rise to a downward sloping curve ($c^A - d^A$), the same as the one already shown in fig. 1(b) = 1(d) = 1(g), because the marginal damage is then assumed to be increasing.

Figure 1

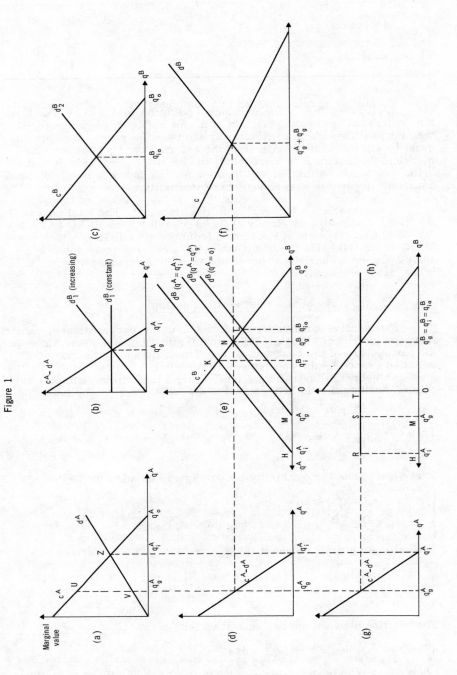

	A	B
i) d^B increasing	$q_g^A < q_i^A$	$q_i^B < q_g^B < q_{ia}^B$
ii) d^B constant	$q_g^A < q_i^A$	$q_g^B = q_i^B = q_{ia}^B$

In all cases the "global optimum" pollution in A will be less than the "individual optimum" pollution in A. As far as B is concerned, if the marginal damage is constant over the relevant range, the optimum quantity will not change because of the externality. If, however, the marginal damage to B is increasing, the pollution created in B in the "global optimum" situation will be intermediate between the "individual optimum" quantities with and without externality.

In no case is the externality fully eliminated. The total cost to B due to the externality is simply reduced from the area HKLO to the area MNLO in fig. 1(e), and it is reduced from the area HRTO to the area MSTO in fig. 1(h). On the other hand, the global optimum solution involves placing an additional burden on country A equal to the area UVZ in fig. 1(a).

A.2.4. Case of two countries: an example

The relative magnitudes of all the changes just mentioned, ranging from the individual to the global optimum situation, are determined by the parameters of the cost and damage functions and by the potential pollution caused by the economic activities in the two countries. Let us give an algebraic illustration of this, using a particularly simple example. Assume that in both countries the functions have the following form: $C = \frac{a}{2}(q_o - q)^2$; $D = bq$

Then $c^A = c^B = a(q_o - q)$
$d^A = d^B = b$

These conditions for optimality become (given an additive TFP):

$$\begin{cases} c^A = a(q_o^A - q^A) = 2b \\ c^B = a(q_o^B - q^B) = b \end{cases}$$

The optimum quantities are then:

$q_g^A = q_o^A - 2b/a$
$q_g^B = q_o^B - b/a$

The relation between the two quantities is:

$q_g^A \gtrless q_g^B$ when $q_o^A \gtrless q_o^B + b/a$

In all cases the absolute amount of pollution abatement will be double in A : $q_o^A - q^A = 2b/a$; $q_o^B - q^B = b/a$.

Now the example just given, which assumes equal damage functions and equal functions for the cost of pollution abatement in A and in B, can represent the situation of two countries or two regions having equal environmental endowments and equal socio-economic structures (but not necessarily the same incomes or populations, although probably having the same per capita incomes and the same patterns of economic activity). Assuming furthermore that $q_o^A = q_o^B$, i. e. that the two countries also have equal incomes and populations, we find a definite relationship between the residual pollution created in A and in B in the global optimum situation, namely: $q_g^A = q_g^B - b/a$. The difference between the two quantities is greater, the higher is the marginal damage of pollution and the lower is the rate of increase in the marginal cost. Needless to say, the total quantity of pollution in B will be greater than in A : $q^T = q^A + q^B = 2q^A + b/a$. The case is illustrated in fig. 2. With reference to a subject discussed later, it is worth while to note that no "equitable" division of the rights of waste disposal seems to lead to the global optimum solution. We have to rely on instruments based on the efficiency principle, which we shall now discuss.

Figure 2

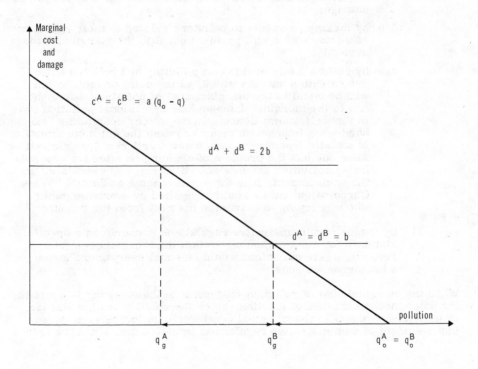

A. 3. Implementation of the global optimum solution

After considering the conditions for optimality, we have to discuss how to implement them. The problem will be examined with reference to the case of two countries with additive TFP. It is well known that the curves for the marginal cost - c^A and c^B - are derived from the horizontal sum of all the marginal curves for the individual polluters, and that the curves for the marginal damage - d^A and d^B - are derived from the vertical sum of the individual marginal damage curves. This means that pollution is a private good from the standpoint of its production or elimination, but is a public good - or a public bad if we prefer to call it that - from the standpoint of its consumption.

The way to reach the optimum solution for TFP must therefore be studied together with the way to deal with pollution within national boundaries.

A. 3.1. Let us start by considering the situation of A before taking account of TFP, so that the "individual autonomous optimum solution" becomes relevant. It is already well known that the optimum quantity of pollution, q_i^A, can be reached:

1) by means of economic incentives:

   1.a by putting a unit tax on waste emissions equal to the marginal damage;

   1.b by making payments to polluters related to their reduction of waste, with a unit payment equal to the marginal damage avoided;

   1.c by putting a bilateral tax on polluting and polluted enterprises with a unit tax which, at the point of equilibrium will be equal to the marginal cost of pollution abatement and to the marginal damage done by residual pollution (this is the well-known Coase scheme which, despite the "excess burden" it imposes in order to reach the optimum situation in a static framework, has some merit in a dynamic setting, when one has the problem of deciding on sites for new polluting activities and new activities exposed to pollution in the environment. It is the same scheme as Scott's "Swiss Corporation" or as would be applied by a private owner who was trying to maximize his rent from the resource;[1]

2) by means of administrative regulations prescribing a specific standard for each emission, so that the marginal cost of reducing waste emissions would be equal everywhere inside a homogeneous zone.

When the marginal cost of pollution abatement in the economy is constant, the price mechanism cannot function, since the solution (with a unit tax or unit subsidy equal to the marginal cost) would be indeterminate, so that a system of administrative regulations is called for. When the mar-

---

1. See R.H. Coase, "The Problem of Social Cost", Journal of Law and Economics, III, pp.1-44 (1960). See also A.D. Scott, op. cit.

ginal cost is increasing (as it will be assumed to be from now on), all the instruments are available. The relevant factors in making a choice will then be:

i) the implementation costs (costs of information, decision-making and control);

ii) the distributive considerations and, more generally, the reactions of the people who influence the political process.

As far as the implementation costs are concerned, it is well known that for the tax system one requires to know only the marginal damage function and the quantity of residual emission; for the payment system one requires to know the marginal damage function and the actual reduction effected in the potential pollution (which is very difficult to find out, in addition to which this system can lead to paradoxes in a dynamic setting, like paying a firm to induce it not to enter the market); and for the administrative regulations system, if it is to be correctly applied, one requires to have a full knowledge of the individual abatement cost functions.

As far as the distributional issues are concerned, it is obvious that the burdens imposed on the polluters and on the victims of pollution will be different under the tax system to what they will be under the payment system. It is also natural that, once the PPP has been accepted, the polluters will prefer the emission standards system to the tax system; while under the first they only have to bear the actual cost of pollution abatement, under the second they have the additional burden of the tax. [1]

The implementation costs and distributional implications help to explain why, notwithstanding the theoretical superiority of the tax system, we find that the administrative regulations system is more widely adopted, together with some subsidization of pollution control measures. Frequently, in order to minimize the implementation costs, a sub-optimum administrative regulations system is adopted which sets the same standard for all discharges, despite the differences in the real cost of pollution abatement.

A.3.2. Taking now the question of TFP, it is easily shown that all the systems available for solving the national pollution problem are also available for solving the international one.

This is obviously true of country B, since the problem of controlling pollution remains a national problem, although it is created by pollution which comes over the frontier (in fact, in the optimum situation we have $c^B = d^B$). But it is also true of country A:

_____

1. It is well known that, in regard to the main distributional issue, i.e. whether to pay compensation to the victims of pollution, the interests of efficiency point to only one conclusion, namely that one should pay the compensation, if a tax is imposed on the polluters and if bargaining with the victims is possible: Otherwise the victims would find it worth while to bargain in order to reduce pollution still further (and therefore more than optimally). In all other instances, whether compensation is paid or not will not affect the optimum solution. See R. Turvey "On the Divergences Between Social Cost and Private Cost", Economica, pp. 309-313 (1963).

a) under the administrative regulations system, all that is required is to impose more severe standards for the emissions in A;

b) under the economic systems, all that is required is to change the unit tax or subsidy, as follows:

tax system	without taking TFP into account	taking TFP into account
$T = t \, q^{A \; 1}$	$t = d^A(q_i^A) = c^A(q_o^A - q_i^A)$	$t = d^A(q_g^A) + m d^B(q^T)$ $= c^A(q_o^A - q_g^A)$
subsidy system $S = s(q_o^A - q^A)$	s = same as t	s = same as t

We can also imagine mixed systems of the following type:

i) A subsidizes the polluters until the individual optimum solution is reached, and B gives them an additional subsidy in order to reach the global optimum solution. In practice, the system would work this way: A would give its polluters a unit subsidy s_2 sufficient to reach q_g^A (the global optimum situation), but would receive back from B the sum: $s_2 (q_o^A - q_g^A) - s_1 (q_o^A - q_i^A)$. The burden on public funds in the two countries is shown in fig. 3, where the vertically shaded area represents the burden on B's government and the horizontally shaded area represents the burden on A's government.

ii) A puts a unit tax t on the polluters sufficient to reach q_i^A, while B gives the polluters in A a subsidy (s-t) sufficient to reach q_g^A. The system can work, since for the single polluter in A the marginal opportunity cost of additional pollution becomes: t + (s-t) = s. As shown in fig. 4, in the optimum situation A's government would obtain total tax receipts represented by the horizontally shaded area, while B's government would pay the vertically shaded area. In a situation which was ideal in the light of the information available, B could limit its payments to the sum = (s-t) . $(q_i^A - q_g^A)$, i. e. to the part of the area lying between q_i^A and q_g^A.

Note, however, that it would require to assess, for each j polluter

---

1. There is the implicit assumption that the whole of the area in A affected by TFP is a homogeneous zone, so that the same unit tax or unit subsidy will apply to all the discharges. Note in addition that the tax and the subsidy are in practice likely to apply more to waste emissions and waste reduction than to pollution and pollution abatement. Nevertheless, given our assumptions, it is sufficient to make a linear transformation of the parameters in order to switch from one system to the other. Considering for instance, the tax, we have:

$$T = t \, q^A = t \sum_{j=1}^{\mu} q_j^A \; (j = 1, \ldots, n \text{ polluters}) = (t/V)W^A = (t/V) \sum_{j=1}^{\mu} w_j^A$$

$(j = 1, .. n)$, the quantity $q^A_{i,j}$ (with $\sum_{j=1}^{\mu} q^A_{i,j} = q^A_i$) that he would reach under the effect of the tax, and that to assess this could be even more difficult than the already difficult task of assessing the potential pollution $q^A_{o,j}$ (with $\sum_{j=1}^{\mu} q^A_{o,j} = q^A_o$).

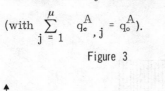

Figure 3   Figure 4

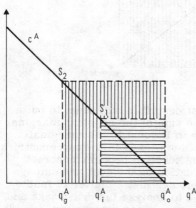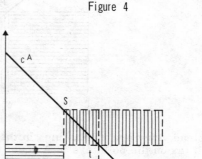

Finally, we can consider the case of the "Swiss Corporation" suggested by Scott. This agency would try to maximize the rent from the resource and would therefore exact from both countries a unit price equal to p in fig. 5, so collecting from A the sum $p\, q^A_g$ and from B the sum $p\, (q^A_o - q^A_g)$, represented respectively by the horizontally shaded area and by the vertically shaded area in fig. 5. [1]

In conclusion, we would repeat that TFP may theoretically be tackled with all the instruments in the wide range which is available for dealing with the problem of national pollution. Nor are the minimum information requirements and the implementation difficulties of the various approaches basically different from those stated for the national case at the beginning of this paragraph.

It is particularly important to remember the differences between the implementation difficulties involved by the optimum solution when the marginal damages in A and in B may be assumed constant and when,

---

1. This result relies on the existence of competitive behaviour on both sides of the market for the resource services. Were the resource owner able to discriminate, he could increase his rent by collecting a surplus from the polluters and the victims. On the other hand, if there were an agreement between the groups of polluters and victims on the rules of the game (for instance, by agreeing to auction one unit of pollution after the other), they could force the resource owner to accept the minimum while keeping all the surplus for themselves. On this subject, see amplius A.D. Scott, "Economic Aspects of Transnational Pollution", this book p. 7.

Figure 5

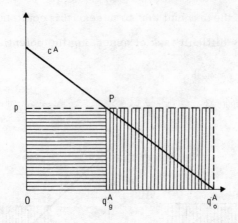

instead, they are increasing. In the first case it is sufficient to put a unit tax on the polluters in A equal to the sum of the constant marginal damages, and a unit tax on the polluters in B equal to the marginal damage in B. In the second case the optimum taxes in the two countries are interdependent of one another and an iterative process becomes necessary (unless we assume full information on the cost and damage curves, so that the optimum solution becomes just a matter of mathematical programming). There is no need to analyse the drawbacks and the real costs associated with the iterative process, even when it tends toward a stable equilibrium, since they are too well known from the literature on the subject.[1] Let us note instead that the problems are, once again, no greater in cases of TFP than in cases of national pollution.

Thus the really relevant difference between international and national pollution problems lies, not in the technical aspects, but in the political ones, being connected with the lack of a central authority which can centralize decisions.

The political aspects have already been outlined in the main paper. Let us now examine, in some of its technical aspects, the major consequence of the lack of a central authority, i.e. the need for bargaining between sovereign nations.

A. 4. Bargaining

The bargaining over one-way TFP concerns, of course, the pollution created in the upstream countries, since internal pollution in the downstream countries has no effect on the upstream ones.

Considering again the case of two countries with the same assimilative environmental capacities it is an acceptable hypothesis that the

1.   See W.J. Baumol and W.E. Oates, "The Use of Standards and Prices for the Protection of the Environment", The Swedish Journal of Economics, No. 1, March 1971; W. Beckerman "Environmental Policy Issues: Real and Fictitious", Problems in Environmental Economics, OECD, Paris, 1972.

starting positions in the bargaining process are determined by the "individual optimum solutions", i.e. for A, the quantity of pollution q_i^A at which $c^A - d^A = 0$, and for B, the quantity q_i^B at which $c^B(q_i^B) = d^B(q_i^T)$ $q_i^T = q_i^A + q_i^B$. When the TFP is non-additive, the initial position for B will be q_{ia}^B, where $c^B(q_{ia}) = d_2^B(q_{ia}^B)$. In this case the "line of maximum resistance" in country B's bargaining strategy is given by:

$$W_o^B = \int_{q_i^A}^{q^A} d_1^B \, dq$$

In order to find the corresponding line when the TFP is of the additive type, we must seek the function which gives, for each degree of pollution, the "minimum total loss" for B, i.e. the sum of the abatement cost and of the residual damage, after B has adjusted optimally to pollution coming from A, i.e. after B has found $q_i^B (q^A)$:

$$L^B(q^A, q_i^B(q^A)) = C^B(q_o^B - q_i^B) \; \Big| \; q^A + D^B(q^T) \; \Big| \; q^T = q^A + q_i^B (q^A)$$

Differentiating in respect of q^A :

$$\frac{d \, L^B}{dq^A} = \frac{\partial L^B}{\partial q^A} + \frac{\partial L^B}{\partial q_i^B} \Big/ \frac{dq_i^B}{dq^A}$$

when the marginal damage in B is constant, $dq_i^B / dq^A = 0$ (as shown in Fig. 1. h), so that dL^B/dq^A is equal to the constant d^B. When instead d^B is increasing, $dq_i^B / dq^A < 0$ and dL^B / dq^A becomes variable, i.e. it decreases with decreasing values of q^A.[1]

Starting from q_i^A and bargaining for a smaller q^A, the line of maximum resistance for B is then given by:

$$W_o^B = \int_{q_i^A}^{q^A} \frac{dL^B}{dq} \, dq$$

If d^B is a constant ($d^B = k$), w_o^B becomes a linear function $= k (q_i^A - q^A)$.

---

1. By diminishing q^A by one unit, B would gain, _ceteris paribus_, the marginal damaged avoided d^B; but B finds it worth while to increase the internally created pollution q^B, so reducing the gain from the damage avoided, but obtaining a saving in the cost of pollution abatement. Given an increasing cost of pollution abatement and therefore a decreasing saving from reducing the abatement level, the net marginal gain from decreases in q^A will also decrease.

For A, the corresponding line of maximum resistance - if we do not introduce other economic and political pressures into the picture - is given by the integral of the net marginal cost of pollution abatement, i.e. the marginal cost less the marginal damage avoided: [1]

$$w_o^A = \int_{q_i^A}^{q^A} (c^A - d^A)dq$$

Fig. 6.a shows the relevant marginal curves and fig. 6.b the corresponding integral curves.

Now, if the pollution abatement operations were normal operations, we would say:

i) that the lines w_o^A and w_o^B would delimit the area in which the agreed solution is bound to lie;

ii) that the solution would depend on the relative strength and the strategies of the parties, as in the normal theory of bargaining. For instance, if the rule of thumb were accepted that "B pays A a unit price for pollution abatement equal to the marginal cost" - which is nothing else than the subsidization method already considered - B would then consider the line $s = (c^A - d^A) (q_i^A - q^A)$.[2] Suppose further that $d^B = k$, so that the linear w^B comes into play. B would then choose the point S (corresponding to the global optimum solution), if it behaved as a perfect competitor, i.e. by choosing the quantity for which $d^B = p$. On the other hand it would choose the point Z, for which $d^B = p + q\frac{dp}{dq}$ if it behaved as a monopsonist.

However, in the special case of pollution abatement, we assume that A will be ready to forego any surplus. This important assumption seems to us quite realistic when we remember that the international community is increasingly assimilating the PPP so that A will at least be induced to "play fair" with B in order to avoid moral condemnation. The assumption implies that A will not try to go beyond w_o^A in fig. 6.b,

so that B could easily reach the optimum point V at which w_o^A is tangent to a parallel of w_o^B, such that $c^A - d^A = d^B$.[3]

1. Note that c^A represents the marginal cost of pollution abatement in A on the following assumptions: (i) pollution is abated optimally in the economy, i.e. by equalizing the marginal cost in all the polluting units; (ii) only the _real_ cost of pollution abatement, in terms of alternative uses foregone, is considered (not including the monetary transfer involved in taxing or subsidizing the polluters). If pollution in A were regulated by imposing equal standards on the emissions (so that the total cost was not minimized), it is obvious that the relevant marginal cost of pollution abatement for the economy would be higher than the c^A which has been assumed so far. Therefore, to say that the maximum resistance line for A is equal to the integral of c^A from q_i^A to q^A ($< q_i^A$) is tantamount to saying that country A is able to regulate pollution optimally and that it takes only the real cost of pollution abatement into account.

2. In the case, shown in Fig. 6.a, of a linearly increasing marginal cost of pollution abatement, s would be twice as high as w^A for any value of q_o.

3. Note that in this case there is no longer a distinction between competitive and monopsonostic behaviour on the part of B. When B has to move along w_o^A, it will always find it desirable to move to a point such as V.

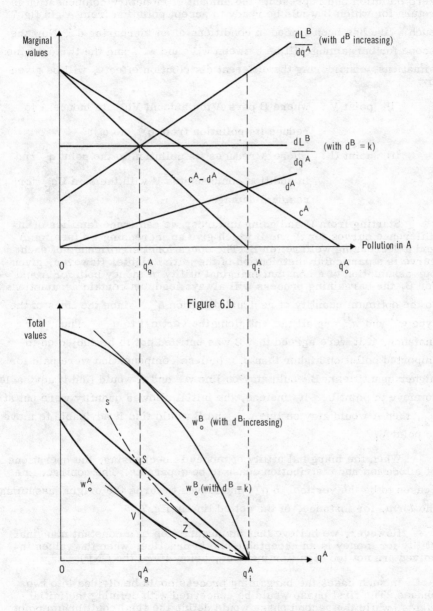

Figure 6.a

Marginal values

$\dfrac{dL^B}{dq^A}$ (with d^B increasing)

$\dfrac{dL^B}{dq^A}$ (with $d^B = k$)

$c^A - d^A$

d^A

c^A

Pollution in A

0   q_g^A q_i^A q_o^A

Figure 6.b

Total values

w_o^B (with d^B increasing)

w_o^A w_o^B (with $d^B = k$)

V   Z   S

0   q_g^A q_i^A q^A

So far so good, but the point is that B is not at all satisfied with the solution, since it still considers itself badly treated. B would be fully satisfied only if it could stay on the line w_1^B which starts from zero pollution and represents the amount of monetary compensation in return for which B would be ready to accept pollution from A. In fig. 7 such a line has been traced in conditions of an increasing d^B. Now the scope for bargaining will be between w_o^A and w_1^B, and the two extreme situations, considering the different distribution effects, will be given by:

i)   point V : where B pays A the amount Vq_g^A to induce A to

reduce its pollution from q_i^A to q_g^A;

ii)   point U̇ : where A reduces its pollution to the point q_g^A and

in addition compensates B with the sum Uq_g^A for

residual damage.

Starting from U and going upwards, we can trace families of in-difference curves for A and B which give an increasing welfare for A and a decreasing welfare for B. The operational significance of each curve is a particular distribution of the initial rights. However, given our assumption of a constant marginal utility of money both for A and for B, the bargaining process will always lead, on certain assumptions, to the optimum quantity of residual pollution q_g^A, since the lines of the type w^A and w^B are all tangent along the vertical to q_g^A. Thus, for instance, if it were agreed that B was entitled not to be subjected to imported pollution higher than q_1^A (unless compensation were paid for higher quantities), B would stay on line w_2^B and A would find it advisable to move to point U_1. If, instead, the initial allowed quantity were put at q_2^A, then A would stay on line w_1^A and B would find it advisable to move to point V_1.

When the marginal utility of money is decreasing, the operations of allocation and distribution cease to be separable. The contract line ceases to be the vertical to q_g^A and moves towards the origin, assuming the form, for instance, of the dotted line in fig. 7.

However, we believe that the assumption of a constant marginal utility for money is an acceptable one in practice, when the values involved are not too high (for instance, below 1% of the GNP).

In such cases the bargaining process could be divided into two phases. The first phase would be concerned with defining the initial rights, while the second phase would define the final equilibrium point to be reached by means of compensation. The first phase would mainly involve a political confrontation, while the second phase could be of a more technical nature. However, that would not make it easier to solve

Figure 7

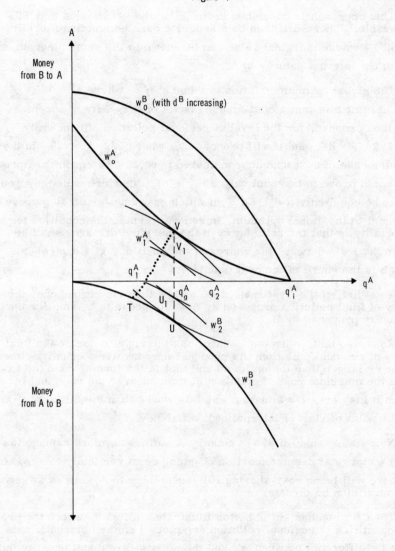

the problem, unless a principle governing the formulation of initial rights could be internationally enforced (which is not likely).

A.5. The cost-sharing approach

The cost-sharing approach (CSA) already analysed above, calls for a few additional notes on the case where the level of pollution is variable (paragraph 4.2.).

This approach is described in fig. 8, on the assumption that TFP is separable. There will then be a separate damage function for TFP - $D_1^B (q^A)$ - whose marginal value can be shown in the same diagram and added to the internal damage in A.

The global optimum solution is found at q_g^A, where $c^A = d^A + d_1^B$. Let us assume now that a cost-sharing basis of fifty-fifty is accepted. A will then propose, for the level of residual pollution, the quantity q_A^A, where $1/2\ c^A = d^A$, and B will propose q_B^A, where $1/2\ c^A = d_1^B$. In the particular case of cost sharing considered here, the two quantities proposed would prove to be equal only if $d_i^B = d^A$, in which case they would prove to be equal only if $d_1^B = d^A$, in which case they would also prove to be equal to the global optimum. In general terms, the condition for such equality is that the ratio between the marginal damages shall be equal to the ratio between the shares of the cost: $d^A / d_1^B = a/b$ (where $a = 1 - b$ is the share of the cost paid by A).

Note that, if the first ratio were higher than the second one, the quantity of final pollution proposed by A would prove to be smaller than the quantity proposed by B.

Note also that, when one country has the right to decide the final quantity of residual pollution, its proposal may involve a quantity either smaller or larger than the optimum and that in the former case the excess of the marginal cost of pollution abatement over the marginal benefit from it may create a situation which is worse than the initial one (at q_i^A) from the point of view of international efficiency.

Note finally that, if at q_i^A country A suffers no more damage (assuming a marginal damage function d^A going down vertically to zero at q_i^A), there will be no cost-sharing rule acceptable to A, unless we introduce compulsion or threats.

The CSA mechanism is better illustrated in fig. 9, where the proposed quantities of residual pollution are plotted on the horizontal axis (so that the difference between q_i^A and the proposed residual quantity indicates the level of pollution abatement) and the share b borne by the victim B is shown on the vertical axis ($0 \leq b \leq 1$). The share borne by the polluter A is of course shown by $(1 - b)$. It is obvious that the welfare of B

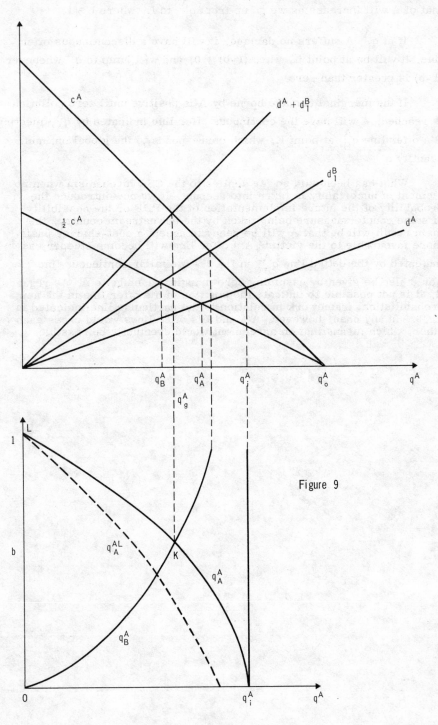

Figure 8

Figure 9

73

will increase as we go down towards zero along his offer line, while that of A will increase as we go up from q_i^A to L, where b = 1.

If at q_i^A A suffers no damage, it will have a discontinuous offer line. It will be at point L, when (1-b) = 0, and will jump to q_i^A whenever (1-b) is greater than zero.

If the marginal damage borne by A is positive until zero pollution is reached, A will have the continuous offer line indicated by q_A^A, meeting B's offer line q_B^A at point K, which corresponds to the global optimum quantity q_i.

What has been said so far applies to the CSA mechanism when no element of uncertainty is taken into account. If we now introduce the probability of the future implementation of the PPP or the possibility of some countermeasure being taken by the downstream country, the main result will be that A will be ready to accept a cost-sharing basis more favourable to the victim. A's offer line will become steeper, as indicated by the dotted line q_A^{A1} in fig. 9. A similar continuous line would also be given by assuming zero marginal damage to A. As regards B, it is not possible to indicate the position of his offer line in the new probabilistical framework in relation to the particular line indicated in fig. 9. In any case, however, A's and B's offer lines would cross each other, which means that an agreement would eventually be possible.

ALTERNATIVE ECONOMIC POLICIES
OF UNIDIRECTIONAL TRANSFRONTIER POLLUTION

by

Henri Smets [1]

Environment Directorate,
OECD, Paris

1.   The opinions expressed in this paper reflect the author's views and not necessarily those of
the OECD Secretariat.

CONTENTS

FOREWORD

The case of one country polluting another country is the simplest example of transfrontier pollution but at the same time it presents a very difficult problem in that the countries are independent of each other. Although the countries can agree on an acceptable pollution level and on efficient instruments to achieve it, each wishes to bear as little cost as possible and thus they have conflicting interests; each country is interested mainly in the maximization of its own national welfare rather than of their collective welfare.

In this paper we have focussed our attention on the distribution issue which is giving rise to the greatest difficulties and we have limited our research to the simple case where the two countries need to find only the proper pollution level and not the level of activities that interact through transfrontier pollution.

After an analysis of the various principles which can be used in solving transfrontier pollution problems, mention is made of the role of international commissions and international standards. This analysis is backed up by a number of theoretical studies presented in the Annexes and its relevance is examined by a detailed consideration of a typical real case of transfrontier pollution. This paper forms a companion to the contribution by Professor Muraro, with whom the author had numerous fruitful discussions during its preparation.

I. INTRODUCTION

Transfrontier pollution, as discussed here, may be defined as pollution in one or more countries due to pollutants generated in another country and transported through natural mechanisms such as air motion and water flow. Cases of transfrontier pollution fall into two categories according to whether one country is polluting other countries (unidirectional transfrontier pollution) or the countries concerned pollute each other (reciprocal transfrontier pollution). Typical cases of unidirectional transfrontier pollution are pollution of international rivers and air pollution from a distant source. Pollution of an international lake or pollution of the airshed of a region extending across a common border are instances of reciprocal transfrontier pollution.

Cases of reciprocal transfrontier pollution can be considered conceptually as the sum of two cases of unidirectional transfrontier pollution occurring simultaneously. Therefore if a general solution to unidirectional transfrontier pollution problems could be found, all problems of transfrontier pollution could in principle be solved.

The purpose of this paper is to investigate whether it is possible to find a uniform and general solution to cases of unidirectional transfrontier pollution. To this end, we will examine the case where pollution is generated only in the "upstream" country and causes damage in the "downstream" country. Clearly the pollution level which maximizes the welfare of both countries will be lower than, or equal to, the pollution level which optimizes the welfare of the "upstream" country, and this "upstream" country would find no direct economic interest in reducing further pollution. The "downstream" country is in a geographically unfavourable position from the point of view of pollution since it cannot pollute the "upstream" country. Thus there is a high degree of asymmetry in the initial positions of the parties concerned.

Problems of unidirectional transfrontier pollution may also be solved by special bilateral arrangements, in which the "upstream" country agrees to keep pollution down as part of a "package deal" covering other issues of a political or economic nature that are not necessarily quantifiable (e.g. defence, diplomatic alliance, trade concessions, cultural agreements, etc.). Such methods of solving the problem will not be examined here because they are of a purely bilateral nature and do not lead to uniform solutions. However, it should be recognized that they are often used and provide a useful way of solving transfrontier pollution problems.

II.   DEFINITION OF THE PROBLEMS TO BE SOLVED

In cases of transfrontier pollution, the countries concerned must first determine the tolerable or optimum level of pollution on the basis of the cost of pollution prevention and control and an estimate of the cost of damage caused by pollution. We assume in this paper that the optimum level is lower than the level which would apply if no pollution prevention and control measures were taken. Secondly they must determine which costs will be borne by each country and in particular whether the "downstream" country will have to pay all or part of the pollution prevention and control costs and/or will receive compensation for damage. Thirdly, some mechanism should be set up to facilitate adjustment over time (e. g. variation of the optimal pollution level). In this paper, we confine ourselves to the case for which the only method available for reducing pollution damage is by abating pollution at the source of emissions, i. e. in the "upstream" country.

III.   PARTICULAR DIFFICULTIES
IN TRANSFRONTIER POLLUTION ISSUES

Transfrontier pollution problems are especially difficult to solve because countries have not yet agreed on the property rights with regard to international water or air and, more specifically, whether a country has initially a right to pollute or to receive a clean medium. Similarly, countries have not yet adopted a principle for the allocation of these rights or the allocation of pollution costs. Internationally-agreed principles of good conduct have not yet been widely adopted and countries have not yet really recognized that their rights to pollute are limited. The present situation corresponds to a transitional period during which countries will have to define rights over the international environment after having defined rights over their national environment. Although these difficulties also arise in national pollution issues, they are much less important because polluters and polluted groups belong to the same community and participate in the same decision-making mechanisms.

IV.   AIM OF THE PAPER

As neither country wishes to create any precedent which would weaken its position in the final bargaining over the international environment, it is likely that seemingly irrational solutions prevail and that the present situation is far from optimal. Thus, the welfare of all countries is reduced until an agreement is reached. The aim of this paper is to discuss various alternatives which could help countries to reach international agreement over transfrontier pollution and thus increase the general welfare more quickly.

We will refer here to the conflict between the "upstream" country and the "downstream" country. In this paper, the term "upstream country" should be understood as a description of the polluters (individuals, firms, municipalities, etc.) and the authorities which deal generally with pollution (authorization, licensing, regulating, monitoring, etc.). The term "downstream" country should be seen as a description of the "victims" of pollution (individuals, firms, public bodies, etc.) of their representative bodies and of authorities which represent the general interests of citizens in this polluted country. No attempt will be made to define which body within each country should cover or bear costs as it is mostly a question of national concern. No implicit assumptions will be made that pollution is "wrong" and that polluted parties are innocent or have a right to compensation.

V. GENERAL ANALYSIS

a) No Damage Estimates Available

In a great number of cases, damage costs cannot be estimated, either because of the lack of suitable information or because pollution affects a public good in the "downstream" countries. In such cases, the "upstream" country will most probably consider that the damage is much smaller than the "downstream" country would claim. Therefore the "upstream" country will not agree to abate pollution to a sufficient extent even if it feels itself responsible for excessive damage.

The only possibility left to the "downstream" country to demonstrate the validity of its claim is to offer to pay the cost of abating pollution to the level which it considers to be tolerable. When the damage is done to a public good in the "downstream" country, it appears necessary to use this solution since the value attributed to this public good can be judged unequivocally only in the light of the political processes in the "downstream" country. If the "downstream" country pays the abatement cost and bears the cost of residual damage, it will presumably be better off than when pollution was not abated. In addition, the welfare of the "upstream" country will remain the same.

Other solutions involving a decrease in the welfare of the "upstream" country do not seem to be realistic as long as the "upstream" country does not accept the damage cost estimate of the "downstream" country. Therefore when damage cost estimates are difficult to establish, the "downstream" countries would probably have to accept the "Victim Pays Principle", at least for some time.

b) Damage Estimates Possible

When the "upstream" country recognizes that transfrontier pollution causes significant damage in the "downstream" country and, at the same time, does not abate pollution, such a choice amounts to behaviour contrary to moral or legal principles proposed internationally and reinforced by the growing concern about excessive pollution in each country. Moral pressure and the danger of creating international friction may thus weaken the position of the "upstream" country.

However, the "upstream" country could claim in some cases that the damage is significant only because the "downstream" country developed pollution-sensitive activities, and that it does not have to bear any liability because of this "wrong" choice by the "downstream" country (just as a car driver might be deemed innocent if a pedestrian jumps in front of the car). Such line of reasoning cannot be used if the "upstream" country obliges a similar polluting activity on its territory to abate pollution because similar pollution-sensitive activities are present in that country. The "upstream" country could also claim that it should not abate pollution just for the sake of allowing the "downstream" country to discharge pollutants or that it should not be subject to more rigorous environmental standards than those applying in the "downstream" country.

This defensive attitude of the "upstream" country is an indication that the "upstream" countries do not feel that they have the right to pollute without giving consideration to the damage in the "downstream" countries, and that they should abate pollution. If the "upstream" country accepts that transfrontier pollution is causing significant damage and that the "downstream" country is acting rationally in the presence of this pollution (e. g. same industry on both sides of the frontier), the "upstream" country should agree to limit pollution at some level which is likely to be higher than that which the "downstream" country hoped for. This would amount to recognition of a limit to pollution which will be a burden on the "upstream" country, especially if it expects to produce more pollutants in the future. The "downstream" country could still obtain a further reduction in the pollution level but would then have to pay for it since the setting of a limit on pollution constitutes an allocation of initial pollution rights. However, economic realities may still be stronger than moral pressure and the "upstream" country may not agree to depart from the "Victim Pays Principle".

In conclusion, this analysis shows that transfrontier pollution problems can be solved either by adopting the "Victim Pays Principle" or by fixing standards of environmental quality on a case-by-case basis. If standards are fixed, the "upstream" country could require that the "downstream" country bears part of the abatement costs, and "downstream" countries could still try to obtain compensation for residual damage. Therefore, in addition to fixing standards, a formula for cost sharing is needed.

The "Polluter Pays Principle" (PPP)

Since relatively little is know about damage costs at present, it might be judged appropriate to avoid that a "downstream" country actually should receive compensation for damage since it would be very difficult to estimate the damage and this would be a reason for delaying any agreement between the countries concerned. Therefore, there would be some merit in adopting a formula by which the "upstream" country pays only all or part of the pollution abatement cost. Such solutions would be a compromise between the "Victim Pays Principle" and the "Civil Liability Principle" which are recommended by the "upstream" and "downstream" countries respectively.

If all countries would accept that the "upstream" country has to pay all the abatement cost, the agreed principle would be the "Polluter Pays Principle" which would then apply not only within each of the 23 OECD countries but also between them. Polluting firms on either side of the frontier would bear the same cost and there would be less chance of distortions in international trade and investment. If the agreed level of pollution is such that there is no residual damage, the PPP is equivalent to the "No Pollution Principle" or the "Civil Liability Principle". When the agreed level is higher, the PPP provides a compromise between the solutions preferred by the two countries.

Agreement on the "Polluter Pays Principle" between countries belonging to an economic community should be easier to obtain than between countries which in their international relationships may make reference to their economic power or military potential. Indeed, the members of an economic community should have due consideration for each other and seek to increase their total welfare without depriving any member of its share of this increase. Furthermore, the acceptance of the "Polluter Pays Principle" within an economic community would have a political meaning, because it implies that the role of intra-community borders would be reduced. If the community forms a geographic bloc that has relatively few cases of transfrontier pollution with the rest of the world, there would be little obstacle to the existence of two principles for transfrontier pollution - one for inside and one for outside the community.

If all countries would accept that the "upstream" country pays, for instance, half of the abatement cost, the agreed principle for transfrontier pollution would be different from the agreed principle for national pollution. However, no distortions would appear in international trade and investment if the transfers from the "downstream" country were considered as a rent to the "upstream" countries rather than to the polluters themselves, and if the polluters in the "upstream" countries had to bear the full abatement cost.

Whichever solution would appear preferable from the outside, the countries will not express their agreement to it until they are ready to agree to a binding compromise. The "Victim Pays Principle" is a recognition of domination resulting from a geographic situation which offers a number of other advantages or disadvantages. Although this solution is not really immoral, there would be some merit in selecting another solution which would go part way to meet the claims by the "downstream" country.

If countries would agree to this intermediate solution, they might find it advantageous to frame the chosen principle in such a way that countries are jointly responsible for total pollution costs (abatement and damage costs). This presentation would avoid stressing the particular liability of the "upstream" country and would also have the merit of including in part the concept of liability for damage. In addition, joint responsibility is a concept that facilitates the choice by each country of the same optimal pollution level.

Multilateral Solution

If no agreement can be reached in the foreseeable future on a single principle applicable to transfrontier pollution, it would be useful to organise

85

a system by which transfrontier pollution problems would be solved without giving official recognition to a principle. In the system proposed in Annex IV, the polluted country would not pay all the abatement cost in the short run but would in fact pay them because of inflation in the long run (just the same as a deposit in the bank is equivalent to a gift of the bank, but is not regarded as such). The polluting country would have an obligation to cover all or part of the abatement cost but it would not see it in this way because this obligation will disappear in the long run because of inflation. If the two countries have to settle two cases of transfrontier pollution in opposite directions and involving the same cost, it is not necessary to specify the liability of each country since the costs to each would be the same with any jointly-agreed formula. The chances of equalizing costs over a large number of transfrontier pollution cases is much greater, just as multilateral payments improve transaction opportunities. In the system described in Annex IV a specialized "bank" would deal with all cases of transfrontier pollution within a certain geographical region such as Europe. This "bank" could also deal with other issues for which costs are clearly defined, such as the construction of a road in one country for the benefit of another country. This system would result in an improvement of the general welfare and could bridge the gap until some rules regarding transfrontier pollution are adopted.

VI.   ROLE OF INTERNATIONAL COMMISSIONS

For each solution, it is necessary to estimate at least roughly, the abatement cost and the damage cost. If each country is tempted to exagerate one of these costs (See Annex I, Section 1(e)), an independent organisation could play a useful role in assessing these costs and in determining the optimal pollution level. Such assessment would not necessarily be correct but it would in principle not be systematically biased in order to satisfy some national interest. This task could be given to a joint commission or an international basin agency provided that it is a really independent organisation. Uniformity within a certain zone would be achieved if there was only one such organisation inside the zone rather than a number of organisations set up by pairs of countries. Uniformity over time would be assumed if such an organisation was a permanent one.

If the initial allocation of pollution rights was made by means of bilateral treaties, the international commission could be assigned the task of finding new pollution levels and new cost allocations which would correspond to new economic situations and be beneficial to the two countries.

If countries would accept a system based on a multilateral settlement of expenditures connected with transfrontier pollution projects (Annex IV), the international commission could act as the technical adviser and the arbitrator in case of disagreement.

VII. INTERNATIONAL STANDARDS FOR AIR AND WATER

A proposal often put forward is that all concerned should agree that international rivers and airsheds should not be subjected to excessive pollution. This approach can be justified on moral grounds and would lead to the establishment of standards which would most probably be too low if they were to be uniform. The acceptance of low standards by all countries would mean a recognition of the right of the "upstream" countries to pollute all media up to this point.

A much more satisfactory solution would be to specify standards for broad categories of situations or for each particular situation by taking into account present and future uses and the extent of pollution abatement being enforced. This solution would involve a large number of bilateral negotiations on standards as well as cost sharing, with the associated danger that no uniform solution would emerge because some countries wield greater economic power than others and could link the negotiations on transfrontier pollution problems with other issues.

VIII. SURVEY OF SOLUTIONS TO REAL CASES
OF TRANSFRONTIER POLLUTION

Many cases of serious transfrontier pollution have been identified only during the last 20 years. Physical measurements have been made but the order of magnitude of the damage cost for the economy is rarely assessed. In a number of instances, transfrontier pollution is decreasing because countries have a national programme for the protection of their own environment or because new technologies are becoming available. However, a few serious cases of transfrontier pollution remain, and they call for urgent consideration.

When problems of unidirectional transfrontier pollution have been solved in the past, it seems that countries negotiated on the basis of some form of reciprocity and there are hardly any examples where two countries officially adopted the "Polluter Pays Principle". Cases where the "upstream" country agreed to maintain, at some cost, the water quality above a certain minimum level without linking this constraint to some other international question are exceptional.

In the case of international common resources, countries have usually shared the cost of joint actions on the basis of the benefit to themselves. This solution does not give recognition to the "Polluter Pays Principle", according to which each polluter should presumably pay in relation to its production of pollutants.

IX. THEORETICAL STUDIES - OUTLINE OF MAIN FINDINGS

In support of the above general analysis, a number of theoretical studies have been carried out with a view to investigating whether there

might be some mutually acceptable solutions to the problems of transfrontier pollution. In these studies, it was assumed that cost functions
were perfectly known at least to one country. The main conclusions of
the studies presented in Annexes I, II and III are given below:

a)   Mutual agreement by countries regarding the acceptable level of
pollution will readily emerge if each country is considered responsible
for a fraction of the total costs associated with pollution. This concept
of joint responsibility derives logically from the fact that transfrontier
pollution is the result of the simultaneous performance of pollution-
emitting activities and pollution-sensitive activities. Joint responsibility
avoids reference to a slogan such as "the polluters must pay" and it can
be organised in such a way as to fit any desired cost-sharing formula
including the one that is included in the "OECD Guiding Principles on
Environmental Policies" under the name "Polluter Pays Principle".
The mutually-desired level of pollution is also the one which maximizes
the total welfare of both countries.

b)   No agreement on the optimal pollution level will emerge automatically if the countries decide to share only the pollution abatement costs.

c)   A number of principles for sharing costs can be used. The "Polluter Pays Principle", for example, can be framed in such a way that
each country is responsible for a fraction of the total cost proportional
to the cost at the optimal pollution level on its own territory (abatement
cost in the "upstream" country, damage cost in the "downstream" country).

d)   In transfrontier pollution issues, the only principle under which
both countries are not worse off is the "Victim Pays Principle". In this
case, the "downstream" country pays all the abatement cost and gains
from the reduction of the total pollution cost. The "upstream" country
does not bear any cost because of transfrontier pollution (and actually
receives a "rent" linked to its geographical position).

e)   Agreements on acceptable pollution levels and cost sharing should
be adapted in the light of changes in economic and social activities and
anti-pollution technologies. Such adaptation will be easier if the "pollution rights" of each country are specified in the agreement.

f)   The current trend in international law, public opinion and international ethics is to recognize that "downstream" countries have some
right to receive air and water of sufficient purity and that "upstream"
countries should take measures designed to avoid excessive pollution.

g)   "Upstream" countries are faced with the danger that other countries may form a coalition and exert pressure in favour of the adoption
of a general agreement on the responsibility of "upstream" countries
in cases of transfrontier pollution. Such a coalition could be quite strong
since only a few countries are "upstream" for all pollution.

h)   In the presence of increasing recognition of a principle of international law assigning responsibility to polluting countries in the case
of transfrontier pollution, "upstream" countries will realize that it is
their long term interest to negotiate an agreement with other countries,
under which pollution costs would be shared between the countries
concerned. (Annex II.)

i)   The concept of uniform income distribution between regions affected by transfrontier pollution does not provide a simple answer to the question of cost sharing. With this ethical concept, the wealthier region has to pay, irrespective of whether it is upstream or downstream.

j)   More interesting results emerge from a cost-sharing approach based on the relative additional income from activities that are the cause of a transfrontier pollution issue. Such "equitable" sharing implies either the "Civil Liability Principle" or the "Victim Pays Principle" if the issue arises from the creation of either a polluting activity in a clean environment or a pollution-sensitive activity in an environment that is already polluted.

k)   The concepts of acquired rights, anteriority and prescription may be used to set up a system which leads to optimum regional planning.

X.   A CASE STUDY

a)   Description

A big river flows from a region (region U) of an "upstream" country to a region (region D) in another country in which there is a lake and then flows into the sea. The big river crosses the border between the two countries but none of its tributaries cross the border. The supply of water for drinking purposes and for industry is provided by tributaries in each country. Water for irrigation has to be taken from the big river. In the past, regions U and D were used for farming by traditional methods. Industrialization of region U brings larger incomes than traditional farming methods while intensive farming using irrigation and tourism in region D bring incomes larger than those from industrialization and much larger than those from traditional farming. In region U it is not possible to develop tourism because of the poor landscape, and irrigation cannot be introduced because the land is not sufficiently flat. In region D, industrialization is possible upstream of the lake but not along the sea where there are other activities.

The industries under consideration for region U emit a pollutant which does not cause significant damage in region U (drinking water comes from an unpolluted tributary and there are no towns or villages near the polluted river). This pollutant may cause serious trouble in region D because it has an adverse effect on agricultural output when irrigation is used and creates disamenities around the lake (odour, foam, etc.), around which tourism can be developed. Pollution abatement techniques are available but they must be used at the point of discharge. The objective of each country is to maximize its income and the only alternatives are traditional farming, industrialization and, in one region, only irrigation and tourism.

b)   Alternatives

Case A: Region U starts industrialization after region D has developed tourism and irrigation.

Case Aa: in region D development of industries brings higher income than further development of tourism and irrigation.

Case Ab: in region D further development of tourism and irrigation would bring higher income than development of industries.

Case B: Region U is industrialized and discharges a pollutant which flows into region D where traditional farming is used.

Case Ba: region D wishes to industrialize.

Case Bb: region D wishes to develop tourism and irrigation.

c) Discussion of the Alternatives

Case Aa

If region U industrialized and does not suffer from pollution, it has no direct interest in bearing anti-pollution costs. The position of region U seems reasonable if the pollution level does not exceed the level which would be acceptable if region D were industrialized like region U. In other words, region U does not consider that its discharge of pollutants is excessive and believes that region D has no "right" to receive pure water even though the water was originally pure. According to region U, region D should not impose obligations on region U because region D decided to develop tourism and region D should cover any anti-pollution costs required for the development of tourism. In its development plans, region D should have considered that region U could only choose industrialization and region D should not have chosen activities which imply obligations on region U without prior agreement on regional planning with region U.

Clearly region D will find it difficult to accept the arguments put forward by region U. First of all, it considers that over time it has acquired a natural "right" to receive unpolluted water and that region U should allow industrialization only if the industries maintain water purity. Secondly, it would point out that similar industries installed in region D would be obliged to take pollution abatement measures and would still make a profit. Finally region D would use a number of arguments based on ethical considerations, internal legislation and international law to claim that transfrontier pollution should not be increased.

Region U would deny that its lag in industrialization opens any right to region D and would point out that industries located in region U would not make a profit if they had to take the anti-pollution measures required in region D because of a number of specific factors which make region U less favourable for industrialization. It would maintain that region D cannot impose obligations which amount to impeding or prohibiting industrialization upstream or force the "upstream" country to stay at the lower income level corresponding to traditional agriculture.

These arguments show that the two regions have opposed interests and viewpoints. However, region D would realize that its choice is between being polluted or paying for pollution abatement if region U does not accept to abate pollution. In other words, region D is better off if it accepts the "Victim Pays Principle" and pays abatement costs as required. If region D believes that agreement on a more favourable principle could eventually come about, it may refuse to pay abatement costs, and thus suffer worse conditions for a while in the hope of ultimately gaining recognition of its right to a clean environment.

Another possibility is for country D to initiate negotiations with country U with a view to having its right to water purity recognized in exchange for some advantage offered to country U. For instance, region D could agree to grant right of passage over its territory for power lines, gas or oil pipelines, or railways, to provide transportation between region U and the sea and built at the expenses of region U. If the benefit to region U of such facilities outweighs the cost of pollution abatement, an agreement on transfrontier pollution and such right of passage would be a preferable solution for region U. This amounts to a reciprocal agreement involving the sovereign rights.

Region D might offer to build a road at its own expense, which would be mainly used for transportation to region U, provided that region U abates pollution. In this case the reciprocal agreement is based on projects that will be of use to the other country, and the multilateral mechanism presented in Annex IV is applicable. If region D cannot offer compensation directly to region U but can offer compensation to another country (e. g. a canal) which in turn can offer compensation to region U (e. g. reduction in transfrontier air pollution), region U may find it useful to abate pollution.

If region U would accept the "Polluter Pays Principle", it might mean that the industries upstream would have to bear high abatement costs to the extent of no longer being competitive. If the rate of profit became too small, industrialization and transfrontier pollution would decrease. Then the "upstream" country could be tempted to provide a subsidy in order to avoid unemployment in region U.

If region U accepted a principle more favourable than the "Polluter Pays Principle" to its industries, the rate of profit of the industries could be sufficient and no subsidy may be needed. However, this could create distortions in international trade and investment because industries might prefer to locate their factories in region U rather than in region D where they have to pay all the abatement costs. Region U could avoid this obvious distortion by requiring its industries to pay the total abatement costs as in region D and by offering to these industries some advantages which eliminate the incentive to switch investment to region D (lower local taxes, cheaper transport rates, etc.).

More generally, the two countries should consider the cost of transfrontier pollution within the broader context of international relations and trade. Although each country would like to obtain public recognition of its claim on pollution costs, even at the expense of offering some hidden advantage, the main objective for each country is to increase its income. If region D creates political friction with region U or causes a reduction of income for region U, it will run the risk of reducing its own income as well, e. g. people living in region U might cease spending their holidays in region D. Similarly, region D is a market for the industrial products of region U. Each country will, therefore, need to act carefully with psychological factors and will probably

91

keep transfrontier pollution problems in perspective. Recourse to the "Polluter Pays Principle" could be a satisfactory solution here.

Case Ab

This case does not differ greatly from case Aa except in that region D cannot show that new industries which are located in region D are profitable even when they pay the abatement costs because there will be no industrialization in region D. Actually, the situation could be that no polluting industries are justified in either region in view of the damage caused to tourism and irrigated agriculture in region D. In such a case, region U will insist that region D should pay the pollution abatement costs because region U does not wish to remain at a low level of income because region D was first to develop tourism and irrigation.

Case Ba

In this case, there is no conflict between the two regions which both become industrialized regions.

Case Bb

Here, region D receives polluted water and is unlikely to develop tourism and irrigation because of poor water quality. If region D becomes industrialized, there is no conflict (Case Ba). However, region D might find that with water at its original quality it could develop tourism and irrigated agriculture. The reasoning would then be the same as in case Aa except that region D has over time "lost" its "natural right" to receive clean water. Thus, region D will find it more difficult to induce region U to abate pollution, and the principle which will probably be applied is the "Victim Pays Principle".

d) Conclusions

The above discussions illustrate the importance of the concepts of acquired rights and income in dealing with transfrontier pollution issues. Whenever pollution costs are commensurate with the additional income derived from activities that give rise to transfrontier pollution issues, the conflict of interests between the countries may be very serious and it will be difficult to find mutually acceptable solutions. If weight is attached to geographical factors (which favour the "upstream" country), the "Victim Pays Principle" would provide a uniform solution. If factors other than the geographical factors are taken into account, other principles could be used but there would still be the conflict of interest because countries need to share the burden of transfrontier pollution and the constraints that this pollution introduces with respect to their economic development. On the other hand, when pollution costs are a minor item in relation to the additional income, difficulties to be overcome in resolving transfrontier pollution issues should not be too numerous. In Annex III, Section 3, we present a detailed theoretical analysis of development strategies in the presence of transfrontier pollution and compare the relative merits of two systems which accept or reject the concept of acquired rights.

e)   Example

In region U, industry produces Q tonnes per year of a good and P tonnes of pollutant ($P = \alpha Q$). The income of industry in region U is βQ. The cost for abating pollution of the river to level p is $\gamma(P-p)$. The income of region D with tourism and irrigation is H in the absence of pollution. Pollution in the river causes a damage evaluated at δHp^2.

The optimal pollution level is found by minimizing

$$\gamma(P-p) + \delta Hp^2 \qquad\qquad (1)$$

and, in this example, is given by

$$P_{opt} = \frac{\gamma}{2\,\delta H} \qquad\qquad (2)$$

if $\dfrac{\gamma}{2\,\delta H} < P$ and is equal to P if $\dfrac{\gamma}{2\,\delta H} \geqq P$.

The cost of abating pollution to the level $\dfrac{\gamma}{2\,\delta H}$ is

$$\gamma\,\alpha\,Q - \frac{\gamma^2}{2\,\delta H} \qquad\qquad (3)$$

and it increases when the activities in region U and D are growing. The damage cost when pollution is at the level $\dfrac{\gamma}{2\,\delta H}$ is

$$\frac{\gamma^2}{4\,\delta H} \qquad\qquad (4)$$

and it decreases when region D develops its activities. When $H \leqq \dfrac{\gamma}{2\,\delta P}$, the damage is δHP^2 and its maximum is

$$\frac{\gamma P}{2}\,(\text{for } H = \frac{\gamma}{2\,\delta P})$$

The total pollution cost at the optimum pollution level is

$$\delta\,\alpha^2 HQ^2 \quad \text{when} \quad HQ \leqq \frac{\gamma}{2\,\delta\alpha} \quad \text{and}$$

$$\gamma\alpha Q - \frac{\gamma^2}{4\,\delta H} \quad \text{when} \quad HQ > \frac{\gamma}{2\,\delta\alpha}$$

The total pollution cost increases with the degree of activities in the two regions.

Figures 1, 2, 3 give the value of the abatement cost, the residual damage cost and the total pollution cost for various values of Q and H.

Case A

In this case, Q is small at the outset and H is large. Region D suffers growing damage when Q increases if there is no pollution abatement. Region D will not offer to pay for pollution abatement until the upstream industry is producing Q* where $Q^* = \dfrac{\gamma}{2\,\delta\alpha H}$ but it might com-

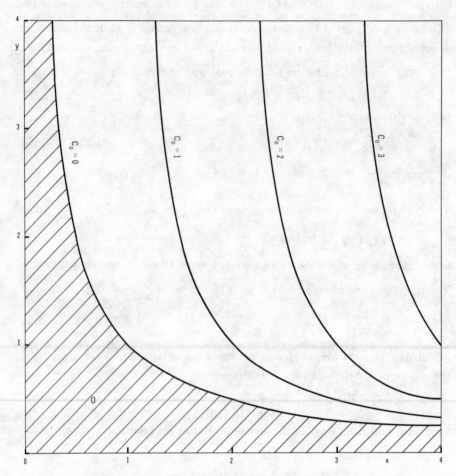

Fig. 1. Optimal abatement cost C_0 as a function of the degree of economic activities in the two countries $\left(x - \dfrac{1}{y} = \text{const. when } xy > 1 \right)$

$x = \gamma \alpha Q$: degree of industrialisation in the upstream country

$y = \dfrac{2\delta H}{\gamma^2}$: degree of development of tourism and irrigation

In the shaded zone, the abatement cost in zero.

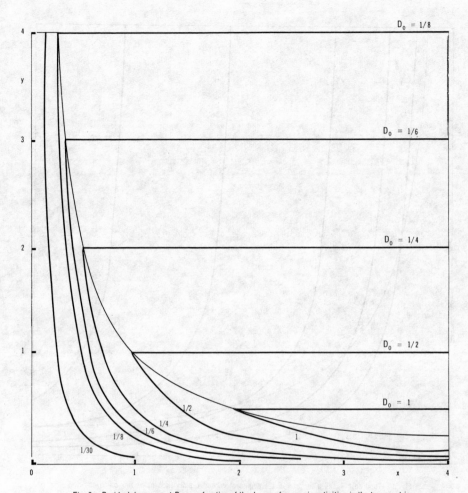

Fig. 2.　Residual damage cost D_0 as a function of the degree of economic activities in the two countries

$(\frac{1}{2y}$ = const. when $xy > 1$ and $\frac{x^2 y}{2}$ = const. when $xy < 1)$

95

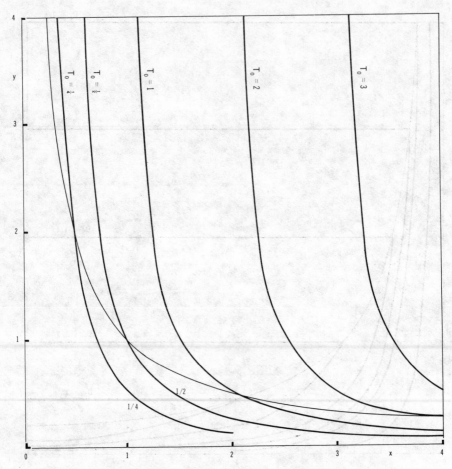

Fig. 3. Total cost T_0 as a function of the degree of economic activities in the two countries

$(x - \dfrac{1}{2y}$ = const. when $xy > 1$ and $\dfrac{x^2 y}{2}$ = const. when $xy < 1$)

96

plain about growing pollution. When $Q > Q*$, region D will probably ask that pollution should not exceed $\dfrac{v}{2\delta H}$. Its income will decrease when Q increases if it has to pay all or part of the abatement cost. However, it will be better off when paying the total abatement cost if this is the only way in which pollution can be maintained at the optimal level.

If a polluting industry is installed in region D, it will have to eliminate all pollution and pay the corresponding abatement cost $vP*$, where $P*$ is the amount of pollutants emitted by this industry. If region U did not industrialize, no abatement cost could have been required if $P* < \dfrac{v}{2\delta H}$. Therefore, industrialization of region U may in fact prevent industrialization of region D.

If region D develops activities on its territory and pays all the abatement cost, it will suffer from increasing pollution cost which may be larger than the added income from these new activities. If region U pays the abatement cost, region D can develop its activities since the residual damage cost is decreasing where H is increasing, but region U will have to bear growing abatement cost. If region D pays half of the abatement cost, the cost due to pollution borne by region D is independent of the level of activities in region D.

Case Bb

In this case, Q is large and H is very small. No pollution abatement takes place and the damage is equal to δHP^2. If the income of region D, $H(1 - \delta P^2)$ is positive, region D could develop its activities. When H reaches the level $\dfrac{v}{2\delta\alpha Q}$, pollution abatement becomes necessary. Region U will seek to avoid paying abatement costs which relate to the activities of region D.

Overall Aspect

The maximum value of the income of the two regions taken together is

$$\beta Q + H - \delta\alpha^2 HQ^2 \qquad \text{when } HQ \leqq \frac{v}{2\delta\alpha}$$

$$(\beta - v\alpha)Q + H\left(1 + \frac{v^2}{4\delta H^2}\right) \qquad \text{when } HQ > \frac{v}{2\delta\alpha}$$

It is a growing function of Q if $\beta > v\alpha$ and a growing function of H for all values of H and Q except for

$$\begin{cases} Q > \dfrac{1}{\sqrt{\delta\alpha^2}} \\ H < \sqrt{\dfrac{v^2}{4\delta}} \end{cases}$$

Figure 4 gives a representation of the total income as a function of the degree of activities in each country.

From a global point of view, it makes no sense to develop the industries to a large extent if $v\alpha > \beta$ and there will be difficulties in developing

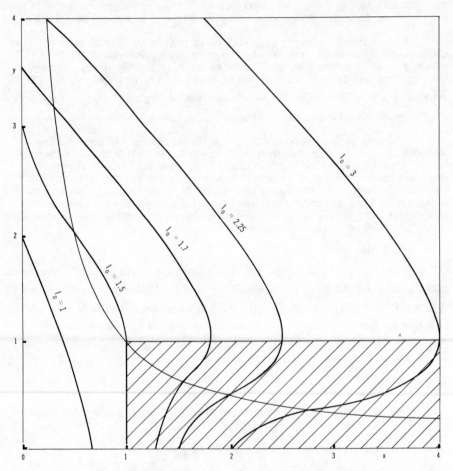

Fig. 4. Total income I_0 as a function of the degree of economic activities in the two countries

for $\frac{\beta}{\gamma a} = 1.5$ and $\frac{\delta^2}{\gamma} = 1$ ($\frac{1}{2} (x + y + \frac{1}{y}) =$ const.

when $xy > 1$ and $\frac{1}{2} (3x + y - x^2y) =$ const.

when $xy < 1$).

The shaded zone corresponds to a decrease in total income when x is constant and y is growing.

98

tourism and irrigation when industrialization has reached a level

$$Q > \frac{1}{\sqrt{\delta \alpha^2}}$$

The first constraint may not be accepted by region U if it does not have to abate pollution. The second constraint could not be found significant to region D if region D activities go rapidly from zero to a level H solution of

$$H + \frac{\gamma^2}{4\delta H} = \gamma \alpha Q$$

which give the same income as when H = 0.

While in this example, the two regions would probably develop their activities as far as possible, in general there is an optimum level of activities in each region and optimum pollution level.

XI.   CONCLUSIONS

In the absence of internationally-adopted rules for transfrontier pollution, countries will most probably have to negotiate bilateral agreements in which transfrontier pollution will be one of several issues. In such agreements, the countries will define the tolerable level of pollution and the "downstream" country will offer some contribution towards the pollution abatement cost borne in the "upstream" country (financial transfer or some advantage on other issues).

If transfrontier pollution is considered separately from other issues, the "upstream" country will insist that the "Victim Pays Principle" should apply or will only accept to limit its pollution at a level which would probably be considered too high by the "downstream" country.

In the presence of political pressure backed up by internationally recognized principles regarding transfrontier pollution, the "upstream" country will abate pollution because of damage in the "downstream" country. The two countries will need to define tolerable levels of pollution based on concepts of economic efficiency, equity and acquired rights. At the same time, they will have to specify which country will pay the required abatement cost and the residual damage cost. Although a general principle could be adopted internationally with respect to cost sharing, the definition of tolerable levels will require a number of bilateral agreements. Among the many possible formulae for cost sharing, special recognition should be given to the principles that imply no compensation for the residual damage in the downstream country. Among these, special attention should be given to the "Polluter Pays Principle" which is the most favourable one for the "downstream" country if residual damage is not to be compensated.

Another approach is based on an explicit reference to damage cost when the two countries can agree on its monetary estimate. In this case, a uniform international rule would consist in fixing the tolerable level of pollution at the level which minimizes the total pollution cost and in making each country responsible for a fraction of the total cost. There are several variants of this method and, in particular, the "Polluter Pays Principle" provides for equality of transfers from one country to the other country. Another variant is examined in the paper and it is shown that it has the merit of being fairly insensitive to exaggerations in cost estimates.

In cases where "upstream" countries do not feel they have any obligation to "downstream" countries and "downstream" countries believe that moral pressure is building up, thus weakening the position of the "upstream" countries, "upstream" and "downstream" countries will often choose solutions which appear to be inefficient in the short run from the economic viewpoint so as to avoid creating precedents that would support the claim of the other country. This stalemate will most probably continue until the countries agree on a compromise of a binding nature which, for instance, could be recognition of the economic efficiency principle and the "Polluter Pays Principle" for cost sharing. In the meantime, they will maintain their opposing claims in order to strengthen their initial position in the final bargaining which would lead to a compromise between the "upstream" country being responsible for all the pollution cost and the "downstream" country being obliged to pay the cost of anti-pollution measures and bear any residual damage cost.

It is therefore concluded that in the absence of moral pressures there cannot be a consensus of views on cost sharing and tolerable pollution levels in the case of unidirectional transfrontier pollution. When moral pressure becomes sufficiently strong, all countries will begin negotiation leading to a uniform rule which would be a mutually satisfactory compromise.

In the meantime, that is, until an international agreement on transfrontier pollution is negotiated, a number of issues relating to transfrontier pollution could be solved without creating precedents by setting up a multilateral system for financing projects related to cases of unidirectional or reciprocal transfrontier pollution. Such a system maintains a balance between the interest of the different countries and it could even be adjusted later on to reflect an internationally-agreed rule which would ultimately be accepted for all cases of transfrontier pollution.

THEORETICAL ANALYSIS OF AN EXTREME CASE
OF TRANSFRONTIER POLLUTION

1. Optimal pollution level and cost sharing principles

In the extreme case of transfrontier pollution, a region in an "upstream" country is polluting a region in a "downstream" country and does not suffer any damage from the pollution in question. It is assumed that pollution abatement is feasible only at the place where the pollutants are discharged, i.e. in the "upstream" country. Let us call the two regions region U (upstream) and region D (downstream).

C(p) denotes the cost of abating pollution in region U to a level p, and D(p) the cost of damage in region D when pollution is at level p in region D.

a) First method

If region U has to bear all or part of the pollution abatement cost:

$$\beta \, (C(p) - B) \tag{1}$$

and region D has to bear the remaining cost:

$$(1 - \beta) \; [C(p) - B] + D(p) + B = C+D - \beta \, (C-B) \tag{2}$$

where β and B are two constants to be determined $(0 \leqq \beta \leqq 1, B \geqq 0)$, region U will not wish to abate pollution whenever $\beta > 0$, or it will not accept to contribute towards the expenses ($\beta = 0$), or it will ask for a contribution from region D $(B = C(\bar{p})$ where $\bar{p}$ is the acceptable pollution level and $\beta > 0$). Any other solution implies that region U would have to bear some cost associated with pollution (e.g. $\beta = 1$, $B = 0$ corresponds to the "Polluter Pays Principle"). Region D will wish the pollution to be abated to level $\bar{p}$, given by

$$(1 - \beta) \left| \frac{\partial C(p)}{\partial p} \right| = \frac{\partial D(p)}{\partial p} \tag{3}$$

In particular if $\beta = 1$, region D will wish the pollution to be reduced to zero since it bears the damage cost and will expect region U to pay all the abatement cost (B = 0).

In general, region U and region D have conflicting interests. Region U would accept to abate pollution only if region D pays for pollution abatement in addition to bearing the damage cost. If region D does not accept this solution and insists that region U should pay all or part of the abatement cost, the two regions will have to negotiate a compromise which will cover both the maximum pollution level $\bar{p}$ and the formula for cost sharing (β and B).

b)   Second method

No disagreement over the maximum pollution level will arise between the two countries if region U is to bear the cost

$$\alpha \: [C(p) + D(p) - A] \tag{4}$$

and region D is to bear the cost

$$(1 - \alpha) \: [C(p) + D(p) - A] + A = C + D - \alpha (C + D - A) \tag{5}$$

where α and A are constants to be determined ($0 \leq \alpha \leq 1$ and $0 \leq A$). If $\alpha = 0$, region U has no preference and region D will pay for all anti-pollution measures.

If $\alpha > 0$, both countries seek a pollution level p_{opt} given by

$$\left| \frac{\partial C}{\partial p} \right| = \frac{\partial D}{\partial p} \tag{6}$$

region U will wish that A be given by

$$A = C(p_{opt}) + D(p_{opt}) \tag{7}$$

and region D will prefer smaller values for A.

With this method, the two countries need only negotiate α and the value of A and not on the pollution level p_{opt}. It thus appears preferable that the polluting and polluted countries should accept to bear costs according to the second method rather than according to the first method. Furthermore, this pollution level is the one which would be chosen when maximizing the total welfare of the two regions taken together.

c)   The Polluter Pays Principle

If the two countries should agree to use the "polluter pays principle" as defined by the OECD, region U would have to bear the cost $C(p_{opt})$, and the constants α and A would be given by

$$\alpha \: [C(p_{opt}) + D(p_{opt}) - A] = C(p_{opt}) \:^{[1]} \tag{8}$$

A suitable solution is $\alpha = 1$ and $A = D(p_{opt})$. Then region U is responsible for

$$C(p) + [D(p) - D(p_{opt})] \tag{9}$$

i.e. anti-pollution cost plus excess damage over the residual damage $D(p_{opt})$. In this case region D bears only the residual damage costs $D(p_{opt})$.

[1].   A more general interpretation of the OECD, PPP could be that $\alpha \: [C(p_a) + D(p_a) - A] = C(p)_a$ where p_a is the agreed or tolerable pollution level not necessarily equal to p_{opt}. This interpretation will not be analysed in this paper.

Another solution is $\alpha = \dfrac{C(p_{opt})}{C(p_{opt}) + D(p_{opt})}$, A = 0.

Region U is then responsible for

$$\frac{C(p_{opt})}{C(p_{opt}) + D(p_{opt})} \; [C(p) + D(p)] \qquad (10)$$

and region D is responsible for

$$\frac{D(p_{opt})}{C(p_{opt}) + D(p_{opt})} \; [C(p) + D(p)] \qquad (11)$$

At the optimal pollution level, region U bears the pollution abatement cost $C(p_{opt})$ and region D the residual damage $D(p_{opt})$. Although this presentation of the "Polluter Pays Principle" does not seem to have been used so far, it has the merit of making the polluting country and the polluted country jointly responsible for pollution effects which occur because of their joint action. The cost to be borne by each country is symmetric and does not lead to interpretation as to whether the upstream country is "guilty" of causing pollution or the downstream country is "guilty" of carrying out activities requiring a pollution-free environment. Such a formula does not incorporate explicitly the concept of "polluter" and "polluted". One country pays in proportion to the cost of abating pollution to the level p_{opt} (or the cost avoided in the absence of a downstream country) and the other pays in proportion to the cost of damage when pollution has reached the level p_{opt} (or the cost avoided in the absence of an upstream country).

Between these two extreme solutions, there are many other suitable solutions which correspond to the definition of the "Polluter Pays Principle":

- optimal pollution level

- pollution abatement cost borne by the polluter

- no compensation for residual damage.

The choice of the largest value of α is justified when it is assumed that the polluter will only react to a charge R if the function to minimize

$$\alpha \, (C + R)$$

is not negligible in relation to other costs. The choice A = 0, i.e. of the smallest value of α, has the merit of minimizing the apparent incidence of transfrontier pollution (which is often aggravated by the fact that "foreign" pollution is regarded as less acceptable than the country's own pollution). In addition, it has the merit of removing the concept of "guilt" and "liability" of one country with regard to the other, and therefore of facilitating the implementation of a rational solution.

d) Other principles

Between the "victim pays principle" favoured by the upstream country and the "civil-party liability principle" requested by the downstream country, the countries could agree on a number of principles for cost sharing. Table 1 lists some alternatives together with the corresponding values for α and A.

Table 1. COST SHARING FORMULAE

	COST TO POLLUTING COUNTRY	COST TO POLLUTED COUNTRY	RELATIONSHIP BETWEEN α AND A	POSSIBLE VALUES FOR α AND A UNDER VARIOUS PRINCIPLES
Civil-liability principle (CLP)	$C_o + D_o$	0	$\alpha = \dfrac{C_o + D_o}{C_o + D_o - A}$	$\alpha = 1$, $\quad A = 0$ (single solution)
1/2 CLP	$\dfrac{C_o + D_o}{2}$	$\dfrac{C_o + D_o}{2}$	$\alpha = \dfrac{\frac{C_o + D_o}{2}}{C_o + D_o - A}$	$\alpha = 1$, $\quad A = \dfrac{C_o+D_o}{2}$ $\alpha = \dfrac{C_o+D_o}{2C_o}$, $A = \dfrac{C_o - D_o}{2}$ ($C_o > D_o$) $\alpha = \dfrac{C_o+D_o}{2D_o}$, $A = \dfrac{D_o - C_o}{D_o}$ ($C_o < D_o$) $\alpha = 1/2$, $\quad A = 0$ (many intermediate solutions)
	D_o	C_o	$\alpha = \dfrac{D_o}{C_o + D_o - A}$	$\alpha = 1$ $\quad A = C_o$ $\alpha = \dfrac{D_o}{C_o+D_o}$, $A = 0$ (many intermediate solutions)
Polluter Pays Principle (PPP)	C_o	D_o	$\alpha = \dfrac{C_o}{C_o + D_o - A}$	$\alpha = 1$, $\quad A = D_o$ $\alpha = \dfrac{C_o}{C_o+D_o}$, $\quad A = 0$ (many intermediate solutions)
1/2 PPP	$\dfrac{C_o}{2}$	$\dfrac{C_o}{2} + D_o$	$\alpha = \dfrac{C_o/2}{C_o + D_o - A}$	$\alpha = 1$, $\quad A = D_o + C_o/2$ $\alpha = \dfrac{C_o}{C_o+D_o}$, $A = \dfrac{D_o + C_o}{2}$ $\alpha = \dfrac{C_o/2}{C_o+D_o}$, $A = 0$ (many intermediate solutions)
Victim Pays Principle (VPP)	0	$C_o + D_o$	$\alpha \, (C_o + D_o - A) = 0$	$\alpha = 0$, $\quad A$ not defined (A=0) or $A = C_o + D_o$, $\quad \alpha$ not defined ($\alpha = 0$)

NOTE: $C_o = C(p_{opt})$, $D_o = D(p_{opt})$.

From a practical standpoint, in many cases little is known about the damage cost and this can be evaluated more accurately by the downstream country. It would, therefore, be preferable that $\alpha < 1$ so that the downstream country will be obliged to reveal its estimate of the optimum pollution level. Furthermore, countries will adopt a more responsible attitude towards achieving minimum cost solutions if they share both the damage cost and the pollution abatement cost (i. e. values of α near 0. 5). In this connection special mention should be made of the symmetric solutions expressing joint responsibility.

Table 2

COST TO POLLUTING COUNTRY	COST TO POLLUTED COUNTRY	PRINCIPLE
$\dfrac{C_o}{C_o + D_o}(C+D)$	$\dfrac{D_o}{C_o + D_o}(C+D)$	PPP
$1/2\ (C+D)$	$1/2\ (C+D)$	$1/2$ CLP [1]
$\dfrac{D_o}{C_o + D_o}(C+D)$	$\dfrac{C_o}{C_o + D_o}(C+D)$	Unnamed (see below section e. 3)

In Annex III, we examine situations in which α is related to the income of the regions under consideration.

Note: If two countries should agree that they have to share responsibility for total pollution cost

region U: $\alpha\ (C+D)$ (12)

region D: $(1 - \alpha)\ (C+D)$ (13)

and try to find a system by which the transfer from region U to region D: $\alpha\,D(p_{opt})$ is equal to the transfer from region D to region U: $(1 - \alpha)\,C(p_{opt})$, they should choose

$$\alpha = \frac{C(p_{opt})}{C(p_{opt}) + D(p_{opt})} \tag{14}$$

In other words, the polluting country would have to bear the cost of pollution abatement and the polluted country the cost of residual damage (Polluter Pays Principle).

e)   Problem of biased cost estimates

e.1) If region D exaggerates the damage $[(1+b)D(p)$ instead of $D(p)]$ and if the abatement cost is known (cost of a treatment plant), region D

1. For a discussion of this principle, see S. C. Kolm: "Guidelines for the Guidelines on the Question of Physical International Externalities", this book p. 249.

105

will ask for a pollution level p_1 which is lower than p_{opt}. If the two regions adopt the Polluter Pays Principle and if region U decides to transfer

$$\alpha_1 [C+(1+b)D] = \frac{C(p_1)}{C(p_1)+(1+b)D(p_1)} \; [C(p)+(1+b)D(p)] \qquad (15)$$

to region D and offers to region D to operate the treatment plant in accordance with the damage in region D, region D would minimize

$$C + D - \alpha_1 \; [C+(1+b)D] = (1- \alpha_1)(C+D) - \alpha_1 bD \qquad (16)$$

The optimum pollution level is $p_2 > p_{opt}$ for which region D maximizes its welfare. When region U discovers that $p_2 > p_1$, it will ask for its contribution to be reduced to $C(p_1)$ and will have an idea of the magnitude of the factor b, of the real damage and of the correct value for the coefficient α . If region D prefers to hide its exaggeration of the damage, its net cost is $D(p_1)$ where $D(p_1) < D(p_{opt}) < D(p_2)$. Therefore the "downstream" country will find it advantageous to exaggerate the damage and to act as though this this damage had been correctly assessed.

e.2) Similarly, if the two countries agree on the use of the "Polluter Pays Principle" and on the damage cost, and if region U exaggerates the pollution abatement cost $[(1+d)C$ instead of $C]$, region U will request a pollution level $p_3 > p_{opt}$. Region D should transfer $(1- \alpha_3)(1+d)C(p)$ - $\alpha_3 D$ to region U, which would apply pollution control measures.
Region U will try to minimize

$$C-(1- \alpha_3)(1+d)C + \alpha_3 D = \alpha_3(C+D)-(1- \alpha_3)dC \qquad (17)$$

This function is at a minimum for $p = p_4$ where $p_4 < p_{opt} < p_3$. If region U reduces pollution to p_4, region D will realize that the pollution abatement cost was exaggerated. It will ask for a reduction in the transfer, and it will know the magnitude of d and the correct abatement cost. If region U reduces pollution to $p_3 > p_4$, the cost will be $C(p_3)$ where $C(p_3) < C(p_{opt}) < C(p_4)$. Therefore, region U would find it advantageous to exaggerate the pollution abatement cost and to act as if this cost had been correctly assessed.

These two examples show that while the two countries may have agreed on the cost-sharing formula, they need to establish some mechanism for the unbiased assessment of the costs (abatement cost and damage cost).

e.3) More generally, if region D says that the damage cost is $(1+b)D(p)$ instead of $D(p)$ and region U says that the abatement cost is $(1+d)C(p)$ instead of $C(p)$, and if the two regions share responsibility:

Region U: $\qquad \alpha [(1+d)C(p)+(1+b)D(p)]$ \qquad\qquad (18)

Region D: $\quad (1- \alpha) [(1+d)C(p)+(1+b)D(p)]$ \qquad\qquad (19)

they will choose the same pollution level p* which minimizes

$$(1+d)C(p)+(1+b)D(p) \qquad (20)$$

They would not choose another pollution level because this would reveal that the declared costs are exaggerated. The level $p*$ is equal to the optimal level p_{opt} when $b = d$. The cost-sharing ratio α is either a constant or a function of $(1+d)C(p)$ and $(1+b)D(p)$.

The net cost to each region is:

Region U: $\quad \alpha \left[(1+d)C(p*) + (1+b)D(p*)\right] - dC(p*)$ (21)

Region D: $\quad (1-\alpha) \left[(1+d)C(p*) + (1+b)D(p*)\right] - bD(p*)$ (22)

and the total cost to both regions is:

$$C(p*) + D(p*) \geq C(p_{opt}) + D(p_{opt})$$ (23)

The net transfer from one region to the other is:

$\alpha(1+b)D(p*) - (1-\alpha)(1+d)C(p*) =$

$\alpha \left[(1+b)D(p*) + (1+d)C(p*)\right] - (1+d)C(p*).$ (24)

This transfer is a growing function of $p*$ when α is constant.

If $\quad \alpha = \dfrac{(1+d)C(p*)}{(1+d)C(p*) + (1+b)D(p*)},$ (25)

there is no transfer and the net cost to region U is $C(p*)$ and the net cost to region D is $D(p*)$ (Polluter Pays Principle). Region U would prefer $p*$ to be large and would choose a large value for d. Region D would prefer $p*$ to be small and would choose a large value for b. Hence both regions would exaggerate their costs and there is no intrinsic limit to such exaggeration. From an efficiency point of view, it is preferable that $b = d$ and it would thus be preferable that each region should have equal possibility of exaggerating its cost.

The cases in which α is not given by Equ (25) are more interesting in that they can lead to situations where both regions do not find any great advantage in declaring costs which are very different from the real costs.

If $\quad \alpha = \dfrac{(1+b)D(p*)}{(1+b)D(p*) + (1+d)C(p*)},$ (26)

the net costs to each region are:

Region U: $(1+b)D(p*) - dC(p*)$ (27)

Region D: $(1+d)C(p*) - bD(p*)$ (28)

If $b = 0$, and $C(p_{max}) = 0$, the net cost to region U takes the values $C(0)+D(0)$, $D(p_{opt})$, $D(p_{max})$ when d takes the values -1, 0 and $+\infty$. The net cost must have a minimum because $C(0) + D(0) > D(p_{opt}) + C(p_{opt}) > D(p_{opt}) < D(p_{max})$. If $d = 0$, and $D(0) = 0$, the net cost to region D takes the values

$\quad D(p_{max})$, $C(p_{opt})$, $C(0)$

when b takes the values -1, 0 and $+\infty$. The net cost must have a minimum because $D(p_{max}) = D(p_{max}) + C(p_{max}) > D(p_{opt}) + C(p_{opt}) > C(p_{opt}) < C(0)$.

From these inequalities, it is reasonable to assume that each region would find that there is a best and finite value for the co-efficients b and d.

107

Generally speaking, it is difficult to make exact calculations because p* is a function of b and d.

Example

If $C = a(P-p)$, $D = up^2$, the optimal pollution level is $p_{opt} = \dfrac{a}{2u}$ provided that $\dfrac{a}{2} < up$.

The pollution level agreed upon is

$$p* = \frac{(1+d)}{(1+b)}\ \frac{a}{2u} \tag{29}$$

provided that $p* < P$. We limit our considerations to values of b and d such that $(1+d)a < (1+b)\,2uP$.

The net cost to region D is

$$(1+d)\,aP - \frac{(1+d)^2 a^2}{4u}\ \frac{2+3b}{(1+b)^2} \tag{30}$$

and is minimum for $b = \dfrac{-1}{3}$. At this value of b, the net cost to region D is

$$(1+d)\,a\!\left[P - \frac{(1+d)a}{2u}\ \frac{9}{8}\right] \tag{31}$$

instead of $(1+d)\,a\!\left[P - \dfrac{(1+d)a}{2u}\right]$ when $b = 0$.

The net cost increases with d when $(1+d) < \dfrac{8}{9}\ \dfrac{uP}{a}$ and decreases when $\dfrac{8}{9}\ \dfrac{uP}{a} < 1+d < \dfrac{12}{9}\ \dfrac{uP}{a}$ \hfill (32)

The upper limit corresponds to $p* = P$. In the worst situation $((1+d)a = \dfrac{8uP}{9})$, the net cost to region D is $\dfrac{4}{9}uP^2$, while if $b = d = 0$ it would be $1/2\ uP^2$ if $a = \dfrac{8uP}{9}$.

The net cost to region U is

$$\frac{a^2}{4u(1+b)}\ (1+4d+3d^2) - daP \tag{33}$$

and is minimum for

$$d = \frac{2}{3}\Big(\frac{uP(1+b)}{a} - 1\Big) \tag{34}$$

i. e.

$$1+d = \frac{2}{3}\ \frac{Pu(1+b)}{a} + \frac{1}{3} > \frac{1}{3} \tag{35}$$

For this value of d, the net cost to region U is

$$\frac{a^2}{4u(1+b)}\ \left[1 - 3\frac{4}{9}\ \Big(\frac{uP(1+b)}{a} - 1\Big)^2\right] =$$

$$-\frac{1}{3}\ \frac{a^2}{4u(1+b)} + \frac{2aP}{3} - \frac{1}{3}u(1+b)P^2 \tag{36}$$

108

and it is always smaller than $\frac{a^2}{4u(1+b)}$, the net cost for d = 0. This cost decreases when b is growing and $u(1+b)P > \frac{a}{4}$ is larger when b is negative. If $b = -\frac{1}{3}$, the net cost is given by

$$-\frac{a^2}{8u} + \frac{2aP}{3} - \frac{2uP^2}{9} \tag{37}$$

It is positive for $0.2 < \frac{uP}{a} < 2.8$, maximum for $\frac{uP}{a} = 1.5$
and larger than $\frac{a^2}{4u}$ for $4/9 < \frac{uP}{a} < 4/3$

Combining the values found for b and d, it is concluded that the above calculations are only valid if

$$p^* = \frac{a}{2u} \left(\frac{1}{2} + \frac{2}{3} \frac{Pu}{a}\right) < P \tag{38}$$

i.e. $uP > \frac{3}{8} a$

If the exact cost is such that $D(P) > \frac{3}{8} C(0)$, the two regions will select values b and d given by

$$\begin{cases} b = \frac{-1}{3} < 0 & (39) \\ \\ d = \frac{2}{3} \left(\frac{uP}{a} \frac{2}{3} - 1\right) & (40) \end{cases}$$

Region U for which the net cost is $\frac{a^2}{4u}$ when b = d = 0
will bear $\frac{-a^2}{8u} + \frac{2uP}{3} - \frac{2uP^2}{9}$ (Fig. 5) $\tag{41}$

Region D for which the net cost is $aP - \frac{a^2}{2u}$ when b = d = 0
will bear $\frac{aP}{6} + \frac{uP^2}{3} - \frac{a^2}{16u}$ (Fig. 6) $\tag{42}$

The total cost to both regions is

$$\frac{5}{6} aP - \frac{3a^2}{16u} + \frac{uP^2}{9} \tag{43}$$

instead of $aP - \frac{a^2}{4u}$ when b = d = 0

Although this solution is sub-optimal (p* $\neq$ P$_{opt}$) the cost for sub-optimality (Fig. 6)

$$\frac{uP^2}{9} - \frac{aP}{6} + \frac{a^2}{16u} \tag{44}$$

is smaller than $\frac{uP^2}{9}$ when $uP > \frac{3}{8} a$. In our example, such a cost is
less than one ninth of the cost of damage on region D when there is no agreement on pollution abatement.

Therefore, acceptance by both regions to bear liability in proportion to the cost in the other country leads automatically, in this example, to a single solution which is mutually acceptable to both regions and there is no cause for one country to argue about the cost estimates made by

109

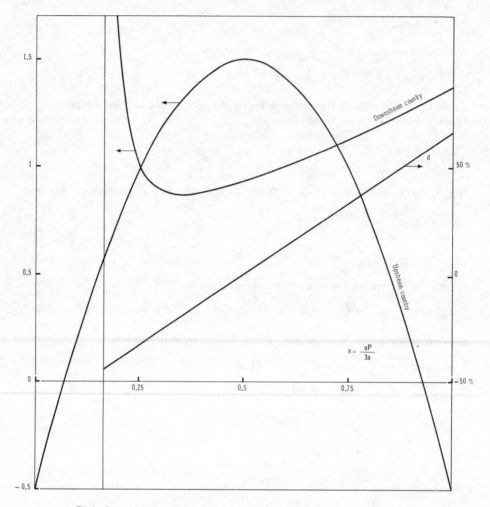

Fig. 5. Ratio of net cost with and without exaggeration for the two countries and of the optimal factor d
$[d = \frac{2}{3}(2x - 1)]$ as a function of $x = \frac{uP}{3a}$. $p_{opt}/P = 1/(6x)$

Ratio of net cost to the upstream country : $8(x - x^2 - \frac{1}{16})$

Ratio of net cost to the downstream country : $\dfrac{6x^2 + x - \frac{1}{8}}{6x - 1}$

The ratios of net cost differ from 1 by less than 50 % when $0.2 < x < 0.85$ (moderate decreases in pollution level)

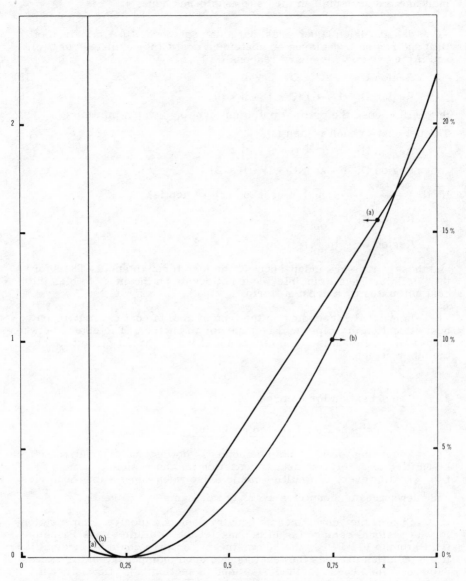

Fig. 6. Curve (a) : Ratio of cost related to suboptimality to the net cost to the upstream country
without exaggeration $(4x^2 - 2x + 1/4)$

Curve (b) : Ratio of cost related to suboptimality to the total cost without exaggeration
$[4x^2 - 2x + 1/4 \; / \; (12x - 1)]$

the other country. Each country will give estimates which are different from the real cost and neither country will be able to tell whether the other has given a correct estimate or not. The result of not giving true values of the cost will be reflected in an increase in the total pollution cost, because the desired level of pollution will not be optimum. This increase is very small in the case considered here.

e.4) If region D declares that a damage cost $D(p) + b$ instead of $D(p)$ and region U declares an abatement cost $C(p) + d$ instead of $C(p)$, and if the two regions share responsibility,

$$\text{Region U: } \alpha \, (C + D + b + d) \tag{45}$$

$$\text{Region D: } (1 - \alpha) \, (C + D + b + d) \tag{46}$$

they will choose the optimal pollution level p_{opt} which minimizes $C + D$. The net cost to each region is

$$\text{Region U: } \alpha \, (C + D + b + d) - d \tag{47}$$

$$\text{Region D: } (1 - \alpha) \, (C + D + B + D) - b \tag{48}$$

If $\alpha = \dfrac{C + d}{C + D + b + d}$, the net cost to each region is

Region U: $C(p_{opt})$

Region D: $D(p_{opt})$

and these values are independent of the bias terms b and d. Therefore the "Polluter Pays Principle" does not create an incentive to bias the cost estimates by a constant term.

In general, depending on the type of bias in cost estimates, one cost allocation principle will be superior to another. The choice between them will depend on the type of biasing which might not be detected by the other region.

2.   Compensation for damage

a)   Compensation between countries

According to the principle adopted, "downstream" countries might be entitled to receive some compensation for the residual damage $D(p_{opt})$. However, difficulties would arise when "upstream" countries and "downstream" countries evaluate this damage differently.

Even if the "downstream" country does not receive compensation for the residual damage (as in the case of the "Polluter Pays Principle" (PPP) or the 1/2 PPP), the "downstream" country could be indemnified if the pollution exceeds the level agreed upon (compensation for excess damage in the second method examined in section 1a). Downstream countries will therefore prefer the second method to the first, even if they both give the same result at the level agreed upon.

b)   Compensation within the downstream country

Enterprises which suffer from transfrontier pollution at an agreed level should not expect to receive compensation when they do not receive

112

any for similar pollution arising in their own country. However, they might receive assistance if the transfrontier pollution level is exceeded since they should not suffer from any failure to keep transfrontier pollution down to the agreed level. In some instances, the government of the downstream country may have agreed upon a higher level in exchange for some other advantages which may or may not be related to its environmental policy. The enterprise suffering from this increase could then receive compensation from the "downstream" government.

3.   Adjustment of the desired transboundary pollution level

In Section 1, it was shown that the acceptable pollution level is chosen in the light of pollution control costs in the "upstream" country and damage costs in the "downstream" country. Both costs can change over time, for the following reasons:

- growth or decline of polluting activities

- improved and cheaper pollution abatement methods

- availability of alternative methods for avoiding or abating pollution

- growth or decline in activities requiring a clean medium

- increased perception of pollution damage

- growth of population exposed to pollution.

Hence, the acceptable level for transfrontier pollution should not be fixed once and for all if economic efficiency is to be maintained.

Let us consider the case where countries have agreed at one point in time on the pollution level p_{opt}.

With such an agreement, two systems can be used:

1)   the agreement specifies only the value of the tolerable pollution level and the amount of the transfer (in respect of pollution control costs) from the "downstream" country;

2)   the agreement specifies the method to be used for choosing the tolerable pollution level and the fraction of pollution control costs borne by the "downstream" country.

With the first system, no difficulty of interpretation will arise as regards the "pollution rights" of the two countries. These rights could be the subject of further negotiation to achieve better economic efficiency if technico-economic conditions should change. With the second system, economic efficiency would be reached immediately but countries could disagree on the cost estimates and initial "rights" are not specified.

Illustrations

First system

If the damage cost increases, p_{opt} should decrease and the abatement cost should increase. If the downstream country asks for a lower p_{opt}, it will have to pay the additional abatement cost so that the upstream country does not have to bear higher costs.

If the abatement cost decreases, p_{opt} should decrease. The upstream country will not accept a decrease in p_{opt} unless the downstream country pays the difference between the abatement costs at the old and the new levels of pollution.

In these two examples, the downstream country will have to work out for itself whether a decrease in pollution level is worthwhile and it will arrive at the most advantageous decision since it has a better knowledge of the true damage cost (but difficulties can arise if the upstream country does not supply correct abatement costs).

Second system

If the damage cost increases the upstream region would have to bear higher abatement costs because p_{opt} decreases. Hence, the upstream region would try to show that the damage costs are exaggerated, in order not to bear higher costs.

If the abatement cost decreases the upstream region would first try to take full advantage of this decrease by not revealing it. When this strategy fails, a new pollution level will be set out and the abatement costs will be higher or lower, depending on the case in question.

In these two examples, the countries will need to refer to some independent institution which will calculate the accepted level of transfrontier pollution in light of its estimates of pollution costs.

4.   Cost sharing in the light of international law

a)   General principles

In the past few years, efforts have been made to define generally the responsibilities of states with regard to transfrontier pollution.[1] The Helsinki Principles of the International Law Association, the Draft Fresh Water Convention prepared within the Council of Europe, and the Declaration on the Environment adopted by the UN Conference in Stockholm may be mentioned in this connection.

One principle which can be followed is to oblige all polluting states to avoid any increase in the level of pollution of media crossing borders. This means that the polluting state has to bear the pollution abatement costs needed to get down to this level or to refrain from some polluting activities, and that it does not have to pay any damage cost when the pollution is held at the present level. When this principle is followed, the optimum pollution level or the tolerable level is fixed at the present level, and the "Polluter Pays Principle" applies. If the polluting state wishes to increase the level it has to "bribe" the polluted state (by paying a compensation for additional damage), and if the polluted state wishes to lower the level it has to pay the additional abatement costs. In practice,

---

1. For a more detailed treatment of this section, see C.B. Bramsen, "Transnational Pollution and International Law", this book, p. 257.

114

this principle amounts to an allocation of pollution rights between the two states on the basis of the prevailing situation.

Another principle is that each state should not cause damage to the environment of other states and should be responsible for any damage caused. This principle implies that the polluting state has to bear all costs (abatement and damage), and that it has no "pollution right".

While the second principle has many points in its favour from the standpoint of ethics, it seems doubtful whether it will be really accepted by upstream states, who will not easily forego their "pollution rights". General acceptance of the first principle would be such a step forward that it would seem reasonable to leave aside this second principle for the time being.

The first principle has the advantage of simplicity since the allocation of "pollution rights" or "pollution franchise" is automatic. However, it could be made more acceptable to all countries if the tolerable levels were agreed upon by the countries concerned in such a way that the agreed levels would be in line with contemplated activities, having due regard to the available pollution control technologies. For instance, the agreed levels would take into account proposed sewage treatment plants and economic development plans of the upstream and downstream regions.

Such deviations from current practice could more readily be agreed upon if the principle underlying the negotiations was to depart from today's practices only by mutual agreement. Short of such agreement, today's pollution levels would be used. It would probably be easier to choose mutually-agreed new pollution levels if there was a jointly managed institution which would examine the benefits and cost of departures from previously accepted levels and of fixing pollution levels in the context of present or contemplated economic development plans.

An international basin agency or an international commission can operate satisfactorily if there is a minimum of agreement between the countries concerned. In the case of a one-way externality, the two countries must first agree on a certain allocation of pollution rights; otherwise the two countries will only transfer the disagreement to the joint institution, which may not be in any better position to resolve it. One significant advantage of such an international institution is that it is able to modify pollution levels in order to reflect economic and technological changes (shut-down of a polluting activity, introduction of a new activity requiring a clean medium, availability of new pollution control techniques at reasonable cost, etc.).

b)   Specific cases

In the case of a number of international rivers, states agreed not to pollute the waters or not to cause excessive pollution of the waters, as part of treaties signed years ago. Since that time, the waters have become heavily polluted and the signatories have not complained about this breach of the agreement. While the upstream signatories would probably argue that the clauses relating to pollution have lost their force, downstream signatories could seek recognition of the validity of the pollution clause by recourse to a competent international court when such a court exists. As downstream countries do not seem to take this course of action, a probable conclusion is that these old treaties do not offer much help in resolving transfrontier pollution problems.

5.   Cost sharing in the presence of social, political or economic pressures

Economic pressure may be used to induce one country to bear some of the costs associated with transfrontier pollution or to reduce such pollution. However, this approach is outside the scope of this study since it is not clear which country can apply the stronger pressure.

In the case of rivers, countries can link problems of transfrontier pollution with those of water management. The downstream country agrees to authorize some work to be undertaken on its own territory (e. g. , a new canal) only if the water quality is restored. Although such a case may appear typical of a reciprocal agreement, it may in fact amount to a situation involving economic pressure (e. g. , when the work authorized is paid for by the requesting country, and the downstream country receives clean water in exchange for an authorization which costs nothing).

Social or political pressures may be applied by the downstream country if it is seriously affected by pollution from the upstream country. Public opinion can be oriented in such a way as to induce the upstream country to abate pollution rather than have a bad public image in the downstream country and possibly in the other countries as well.

Polluted countries can also form a coalition and establish rules of good conduct in transfrontier pollution matters, which would be acceptable to downstream and onstream countries and which would not entirely satisfy the upstream countries. If the countries in favour of these rules wield sufficient political power, upstream countries might in the end have to accept the new rules in view of the pressure of international public opinion and the danger that a coalition of signatories might apply significant economic pressure. Thus, upstream countries cannot afford to disregard transfrontier pollution problems and it is in their interest to seek a solution that will be generally acceptable.

If the upstream countries should prefer not to be bound by formal agreements, they could exercise self-discipline to ensure that their pollution does not cause excessive damage. Such a strategy leads to the de facto determination of a tolerable pollution level. A further analysis of this problem is to be found in Annex II.

Annex II

DYNAMIC CONSTRAINTS IN THE PROBLEM
OF TRANSFRONTIER POLLUTION

INTRODUCTION

In this annex we show, from a few simple cases, what role a change in the internationally adopted rules would play in the problems of transfrontier pollution. Meanwhile a more comprehensive and detailed study is being prepared. A brief formulation of the problem as a whole is given in the Appendix which follows.

Let us take the case of unidirectional pollution in which the upstream country discharges a pollutant which causes damage only in the downstream country. Let us suppose that for technical reasons it is only possible to abate the pollution by processes to be applied at the source, i. e. in the upstream country. Let $D(p)$ be the damage cost; $C(p)$ the cost of the pollution control measures; p the level of pollution and p_{max} the level of maximum pollution (i. e. the level when $C = 0$). Let p* be the optimum level of pollution which minimizes the sum of

$$C(p) + D(p).$$

Let us suppose that the level p* is lower than p_{max} and that the problem of transfrontier pollution will arise at some future time when there is a degree of probability q that the principle of international law recognized by the countries concerned will be the principle of civil liability (i. e. that the upstream country is responsible for the total damage caused by transfrontier pollution).

The probable cost to the upstream country due to transfrontier pollution will then be $qT(p) = q[C(p) + D(p)]$. Since the upstream country will prefer to minimize the cost it has to bear, it will prefer the level p* and the probable cost will then be qT(p*).

First Strategy

If the downstream country adopts a passive and rational attitude, its probable expenditure due to transfrontier pollution will be $(1-q) T(p*)$, since it will prefer to pay $C(p*)$ to the upstream country to reduce the damage for which it is held solely responsible. The upstream and downstream countries might agree on a cost-sharing formula which would remove all uncertainty and they might finance respectively $\alpha T(p)$ and $(1-\alpha) T(p)$ where α is a constant.

117

This agreement would be mutually satisfactory if:

$$\begin{cases} qT(p^*) - \alpha\, T(p) \geqq 0 & (1) \\ (1-q)T(p^*) - (1-\alpha)T(p) \geqq 0 & (2) \end{cases}$$

Each country would derive maximum benefit (or advantage) if $p = p^*$ and conditions (1) and (2) would be satisfied simultaneously only when

$$\alpha = q \qquad (3)$$

Accordingly the two countries might conclude a cost-sharing agreement and substitute known costs for unknown costs. The cost-sharing co-efficient would depend on the probability q and the level of pollution chosen would be the optimum level of pollution.

Second Strategy

A more interesting case is provided by the assumption that the downstream country seeks to increase the probability q by refusing to consider the possibility that it might cover part of the cost of the pollution control measures. In this case it will meet the cost $D(p_{max})$ if the principle of civil liability is not adopted (probability $1-q$), and the cost 0 if this principle is adopted (probability q). The probable cost to the downstream country due to transfrontier pollution then becomes:

$$(1-q)\, D(p_{max}).$$

The probable cost to the upstream country due to transfrontier pollution will be $qT(p) + (1-q)\,0$ in the same circumstances. The upstream country will minimize this cost by putting pollution control measures into effect which enable the level p^* to be reached.

A cost-sharing agreement between the upstream $[\alpha\, T(p)]$ and downstream $[(1-\alpha)\, T(p)]$ countries will be mutually satisfactory if the "benefits" (or advantage) G_U and G_D are not negative.

$$G_U \equiv qT(p^*) - \alpha\, T(p) \geqq 0 \qquad (4)$$

$$G_D \equiv (1-q)\, D(p_{max}) - (1-\alpha)\, T(p) \geqq 0 \qquad (5)$$

Each of the countries will derive the maximum "benefit" if the level of pollution mutually agreed on is p^* and the only point to be negotiated will be the value of the co-efficient α.

The "benefits" to the two countries will be positive if there is at least one value for α which satisfies

$$q - (1-q)\frac{D(p_{max}) - T(p^*)}{T(p^*)} < \alpha < q \qquad (6)$$

As $p_{max} > p^*$ and $D(p_{max}) > T(p^*)$ this condition is always satisfied and α can have any value within the range defined by condition (6).

In considering these various possibilities it is desirable to stress the principle of equality of "benefits" for the two countries. In this case the cost-sharing co-efficient will be given by

$$\alpha = q - \frac{1-q}{2} \left(\frac{D(p_{max}) - T(p^*)}{T(p^*)} \right) \qquad (7)$$

Agreement on this value for α will only be reached if $\alpha > 0$, i.e. if:

$$q > \frac{D(p_{max}) - T(p^*)}{D(p_{max}) + T(p^*)} \qquad (8)$$

The strategy adopted by the downstream country will only be rational if the co-efficient α given by the equation (7) is higher than the probability q when the downstream country is passive (Equation (3)).

In short, it would appear that both countries stand to gain from negotiating a cost-sharing agreement as soon as there is sufficient probability that a rule based on civil liability will be adopted internationally. The two countries might try to influence the value of this probability, thereby adding extra costs to the costs connected with transfrontier pollution and departing still further from the optimum situation. The range of possible values for α becomes wider when the downstream and upstream countries incur expenditure in order to bring forward or postpone the date for adopting a uniform rule, and it also becomes wider if account is taken of the fact that the disutility of a commitment to spend an unknown sum is often deemed greater than when the sum to be spent is known ("risk aversion").

While the upstream country has no direct interest at present in bearing expenditure corresponding to the transfrontier pollution, it will reconsider its position as soon as it realises that, all things considered, a cost-sharing arrangement is to be preferred to a contingent liability to assume full responsibility for damage due to transfrontier pollution. Thus the dynamic constraint imposed by the possibility that the principle of civil liability may be adopted in issues between states (or by polluters and victims of pollution on either side of a frontier) may be advanced as a strong argument for reaching early mutual agreement on a less onerous rule, as for example the rule of equally divided responsibility ($\alpha = 1/2$).

The downstream country will not fail to realise the value of such an agreement, since it will find it easier and cheaper to have a signed treaty put into force rapidly than a rule of international law.

Sharing Pollution Control Costs

Instead of sharing the total costs due to pollution, countries could agree to share only the pollution control costs. The upstream countries would apply pollution control measures and would be paid the proportion $(1 - \beta)$ of their cost by the downstream countries. In this case the agreement would be mutually satisfactory if:

$$G'_U = qT(p^*) - \beta C(p) \geqq 0 \qquad (9)$$

$$G'_D = (1-q) D (p_{max}) - (1 - \beta) C (p) - D(p) \geqq 0 \qquad (10)$$

The upstream country would derive maximum benefit when

$$p = p_{max}, \qquad 0 \leqq \beta \leqq 1$$

and $\beta = 0$ $\qquad 0 \leqq p \leqq p_{max}$

The downstream country would derive maximum benefit only when $\beta = 1$ and $p = 0$. In this case the negotiations would cover both the value of β and the level p.

The sum of the "benefits" to each country

$$qT(p^*) + (1-q) D(p_{max}) - T(p) \tag{11}$$

would be greatest if $p = p^*$. Thus both countries could each enjoy their maximum benefit if they choose the optimum level of pollution p^*.

In the case under consideration there is always a positive value for β which is mutually satisfactory, since the conditions (9) and (10) with $p = p^*$ correspond to

$$qT(p^*) - (1-q) \left[D(p_{max}) - T(p^*) \right] \leqq \beta C(p^*) \leqq qT(p^*) \tag{12}$$

The benefits to both countries will be equal and maximum if $p = p^*$ and if

$$\frac{\beta C(p^*)}{T(p^*)} = q - \frac{1-q}{2} \frac{D(p_{max}) - T(p^*)}{T(p^*)} \tag{13}$$

It will be seen then that the pollution control cost-sharing coefficient is given by

$$\beta = \alpha \frac{T(p^*)}{C(p^*)} \tag{14}$$

when the principle of equality of advantage is adopted. It will depend on the probability q (as modified by the strategies adopted by the countries concerned) and the ratio $C(p^*) \ / \ T(p^*)$.

In the event of the probability q being independent of the ratio $C(p^*)/T(p^*)$, the coefficient β would vary from one case of pollution to another and might even exceed 1 if the residual damage $D(p^*)$ were considerable. Consequently, it could not always be mutually satisfactory to settle on a fixed value for β by international agreement (as for example $\beta = 1$ in line with the Polluter Pays Principle adopted by the OECD for cases of pollution inside a country), but this method has the advantage of not depending explicitly on arriving at an actual figure for damage cost. (Although a national figure will be used in choosing the level p^*.)

Another assumption is that the probability q will vary with the ratio $C(p^*)/T(p^*)$, i. e. that the value of q at a given moment will be greater or less, depending on the damage done in the downstream country and the cost of the pollution control measures in the upstream country:

$$q = q' \frac{C(p^*)}{C(p^*)+D(p^*)}$$

where q' is independent of C/D, but may vary with the strategies adopted.

On this assumption the coefficient α will vary from one case of pollution to another, but the coefficient β will be fixed. In choosing between these two assumptions, one would have to know whether the rule of international law which was likely to be agreed upon was the principle of civil liability or the Polluter Pays Principle.

120

On the latter assumption it would be desirable to examine whether the "polluter pays half" principle (β = 1/2) would not be advantageous, especially if the probability q' was not very great.

CONCLUSIONS

1.   So long as the upstream country does not consider that it might be obliged to pay compensation for damage due to transfrontier pollution, it will have few economic reasons for meeting the cost of abating such pollution. Transfrontier pollution is in the nature of an externality imposed by the upstream country and suffered by the downstream country.

2.   As soon as the upstream country realises that the pollution it discharges is likely to promote the adoption of a rule of international law whereby it would be held responsible for meeting certain expenses in connection with transfrontier pollution, it will agree to negotiate with the downstream country regarding a more favourable cost-sharing arrangement than the legal provisions which would be likely to be introduced. The sooner the upstream country negotiates, the more it will be able to make the basis of the cost-sharing favourable to itself. The cost which the upstream country will agree to bear will depend on the terms of the legal principle likely to be adopted; it will be higher if this is the principle of civil liability and lower if it is the Polluter Pays Principle.

3.   The downstream country will adopt an equitable cost-sharing basis to the extent that it deems it preferable to have a negotiated agreement which can be applied effectively than to put forward claims supported only by international law.

4.   If a basis for cost-sharing is adopted which is generally considered to be fair, the downstream country will be less able to obtain a more favourable basis subsequently. Among the possible bases one should consider (a) equality of total costs (α = 1/2), (b) the Polluter Pays Principle (β = 1), and (c), when the residual damage is slight, the "polluter pays half principle" (β = 1/2).

5.   Ignorance of the real damage costs will make it easier to reach agreement on a value for α or β , in addition to which the absence of additional costs and the removal of uncertainty as to who is responsible are two factors which will make up for the inevitable imperfections in an agreement on a fixed value for the cost-sharing ratio.

6.   In short, dynamic constraints are the basic factor which will promote the general adoption of guiding principles governing transfrontier pollution, even in the extreme case where the upstream country suffers no damage from transfrontier pollution.[1]

1.   a)   The above analysis could have been derived from Nash principle in order to find the optimal solution in a "game" between two countries (Maximization of the product of the "benefits").
        b)   The probabilistic approach in transfrontier pollution problems is also examined by Prof. Muraro and Prof. d'Arge in their papers included in this book.

Appendix

MATHEMATICAL FORMULATION OF THE PROBLEM OF TRANSFRONTIER POLLUTION

Let us first consider the particular case where the effect of pollution control measures taken during the year i is to reduce pollution during that year only and where the damage caused by pollution during the year i depends only on the pollution p_i during that year. The costs of pollution control and the damage costs will vary from one year to another owing to technological change and to variations in the number of polluting activities or of activities sensitive to pollution.

Let

$C_i(p_i)$ be the minimum cost of pollution control during the year i required to achieve a level of pollution p_i during that year;

$D_i(p_i)$ be the value of the damage done during the year i when the pollution level is p_i;

$T_i(p_i) = C_i(p_i) + D_i(p_i)$

$\bar{q}_i \, T_i(p_i)$ be the probable cost estimated and borne by the upstream country due to pollution at level p_i during the year i (N.B. $\bar{q}_i$ is not defined here as the probability that the principle of civil liability will be adopted);

$\bar{A}_i = \bar{q}_i \bar{M}_i$ be the additional cost during the year i borne by the upstream country for the sake of reduing the values of the coefficients $\bar{q}_j$ and q_j $(j \geq i)$ (measures taken to alter these coefficients or in response to similar measures taken by the downstream country). (N.B. $\bar{q}_i$ may be a function of C_i, D_i, M_i and $\bar{M}_i$);

$(1-q_i)T_i(p_i)$ be the probable cost estimated and borne by the downstream country due to pollution at level p_i during the year i;

$A_i = (1-q_i)M_i$ be the additional cost during the year i borne by the downstream country for the sake of increasing the values of the coefficients $\bar{q}_j$ and q_j $(j \geq i)$; (measures taken to alter these coefficients or in response to similar measures taken by the upstream country);

$f(i)$ be the discounting function (monotonic and decreasing);

123

$\alpha_i T_i(p_i)$ be the fraction of the total cost which the upstream country undertakes to bear during the year i;

$\beta_i C_i(p_i)$ be the fraction of the cost of pollution control which the upstream country undertakes to bear during the year i;

$(1 - \alpha_i)T_i(p_i)$

be the fraction of the total cost which the downstream country undertakes to bear during the year i;

$(1 - \beta_i)C_i(p_i) + D_i(p_i)$

be the fraction of the cost of pollution control and the damage cost to be financed by the downstream country during the year i.

If there is no agreement, the upstream country will estimate that the probable discounted cost it will have to bear due to transfrontier pollution will be

$$\Sigma \ \bar{q}_i \ /T_i(p_i) + \bar{M}_i / \ f(i).$$

The probable discounted cost to the downstream country will be

$$\Sigma \ (1 - q_i) \ /T_i(p_i) + M_i / \ f(i).$$

All the values for p_i which minimize these two expressions will depend on the strategies followed by the two countries. If $M_i = \bar{M}_i = 0$ and if q_i and $\bar{q}_i$ are independent of p_i and T_i, p_i will be the value which minimizes $T_i(p_i)$.

If there is an agreement, the upstream country will have to bear

$$\Sigma \ \alpha_i \ T_i(p^*_i) \ f(i)$$

$$\text{or} \ \Sigma \ \ \beta_i C_i(p^{**}_i) f(i)$$

and the downstream country will bear

$$(1 - \alpha_i) \ T_i(p^*_i) f(i)$$

$$\text{or} \ /(1 - \beta_i) \ C_i(p^{**}_i) + D_i(p^{**}_i) / f(i).$$

The sets of values for p^*_i and p^{**}_i will be chosen so as to minimize the costs borne by the two countries. If there is an agreement to share total expenditure, it is clear that p^*_i will be the value which minimizes $T_i(p^*_i)$. In cases where the countries accept the principle of equality of "advantage", p^{**}_i will be the value which minimizes $C_i(p^{**}_i) + D_i(p^{**}_i)$.

Let the following be the average values:

$$\bar{q} = \frac{\Sigma \ \bar{q}_i /T_i(p_i) + \bar{M}_i / \ f(i)}{\Sigma \ /T_i(p_i) + \bar{M}_i / \ f(i)} \qquad q = \frac{\Sigma \ q_i /T_i(p_i) + M_i / \ f(i)}{\Sigma \ /T_i(p_i) + M_i / \ f(i)}$$

$$\overline{T} = \frac{\sum [T_i(p_i) + \overline{M}_i] f(i)}{\sum f(i)} \qquad\qquad T = \frac{\sum [T_i(p_i) + M_i] f(i)}{\sum f(i)}$$

$$N = \sum f(i)$$

$$\alpha = \frac{\sum \alpha_i T_i(p^*_i) f(i)}{\sum T_i(p^*_i) f(i)} \qquad\qquad \beta = \frac{\sum \beta_i C_i(p^{**}_i) f(i)}{\sum C_i(p^{**}_i) f(i)}$$

$$T^* = \frac{\sum T_i(p^*_i) g(i)}{\sum f(i)} \qquad\qquad C^{**} = \frac{\sum C_i(p^{**}_i) f(i)}{\sum f(i)}$$

$$D^{**} = \frac{\sum D_i(p^{**}_i) f(i)}{\sum f(i)}$$

If q_i and $\overline{q}_i$ have increasing values, $\overline{q}$ and q will have values lying between the minimum and the maximum for $\overline{q}_i$ and q_i. One could associate q (and $\overline{q}$) with a probability q_i (or $\overline{q}_j$), i.e. a probability in i (or j) years (a situation in the future).

The benefit to the upstream country from an agreement will be

$$G^*_U = \overline{q}\, \overline{T}\, N - \alpha\, T^*N$$

or $\qquad G^{**}_U = \overline{q}\, \overline{T}\, N - \beta\, C^{**}N$

The benefit to the downstream country from an agreement will be

$$G^*_D = (1-q)\, T\, N - (1-\alpha)\, T^*N$$

or $\qquad G^{**}_D = (1-q)\, T\, N - (1-\beta)\, C^{**}N - D^{**}N$

If one defines the quantity $\widetilde{T}$ as being $\frac{1-q}{1-\overline{q}}\, T$, the benefit to the downstream country will be

$$G^*_D = (1-\overline{q})\, \widetilde{T}\, N - (1-\alpha)\, T^*N$$

$$G^{**}_D = (1-\overline{q})\, \widetilde{T}\, N - (1-\beta)\, C^{**}N - D^{**}N$$

The above equations correspond exactly to equations (1)-(2), (4)-(5), and (9)-(10) given above when dealing with pollution control measures costing $NC(p)$ and pollution damage costing $ND(p)$ at a future time characterized by probabilities q or $\overline{q}$.

Thus it would appear that the highly theoretical case studied in that paper is a condensed expression of the more realistic case presented here.

To be more precise, the above equations correspond to the present case on the assumptions that $q_i = \overline{q}_i$ and that α_i and β_i are constant.

The two strategies are defined by

a) $\qquad M_i = \overline{M}_i = 0$

b) $\qquad \overline{M}_i = 0, \quad M_i = D_i(p_{max.\ i}) - T_i(p^*_i)$

and the characteristic magnitudes are:

$$D(p_{max}) = \frac{\Sigma \; (1-q_i) \; D_i(p_{max,\,i}) \; f(i)}{\Sigma \; (1-q_i) \; T_i(p_i^*) \; f(i)} \; \Sigma_i \; T_i(p_i^*) \, f(i)$$

$$T(p^*) = \Sigma_i \; T_i(p_i^*) f(i)$$

$$C(p^*) = \Sigma_i \; C_i(p_i^*) f(i)$$

$$D(p^*) = \Sigma_i \; D_i(p_i^*) f(i)$$

The general case of transfrontier pollution calls for an examination of the methods of controlling pollution which will bear fruit both in the present and in the future, and of the damage caused by pollution both in the present and in the past. If one knows the present and future cost of the various methods of controlling pollution and how the amount of pollution from each source will vary over time, one can examine various strategies costing $C_s(t)$ during the year t which will achieve levels of pollution $p_s(t)$. The damage resulting from level $p_s(t)$ will be $D(t \; ; \; p_s(t'), \; t' \leqq t)$ and will vary explicitly with time when there are variations in the number of activities affected by pollution and in their sensitivity to pollution.

The overall economic optimum is obtained by finding the function $C_s(t)$ which minimizes

$$\Sigma \; [C_s(t) + D \, (t \; ; \; p(C_s(t')), \; t' \; \leqq \; t) \,] \; f(t).$$

The overall optimum value for the level of pollution during the year t will no longer be obtained by minimizing the sum of the cost of pollution control and the damage cost during that year. In order to calculate the levels of pollution or determine the strategies to be adopted by the upstream and downstream countries in the various hypothetical cases examined above, one has to make a more complex analysis, because one cannot break down the problem into a series of annual problems which can be solved independently.

Annex III

COST SHARING BASED ON INCOME SHARING

1. Comparable economic situation

Pollution abatement costs could be shared in such a way that two regions initially having the same income also have the same income with pollution-sensitive activities in the downstream region and pollution-generating activities in the upstream region, provided that these activities produce maximum income in each region.

Let P_{do} and P_{uo} be the initial income of the downstream and upstream regions, P_{di} and P_{ui} be the income of the two regions in the absence of pollution when they develop pollution-sensitive actitities or pollution-generating activities, D_d the damage costs associated with pollution in the downstream country and C_u the pollution abatement costs in the upstream region. We assume that $P_{di} > P_{do}$, and $P_{ui} > P_{uo}$.

The two regions would remain in a comparable economic situation if

$$\frac{P_{uo}}{P_{do}} = \frac{P_{ui} - \ell\,(D_d + C_u)}{P_{di} - (1-\ell)\,(D_d + C_u)} = \frac{P_{ui} - mC_u}{P_{di} - (1-m)C_u - D_d} = g \qquad (1)$$

where ℓ and m are two constants associated with cost sharing ($0 \leq \ell \leq 1$, $0 \leq m \leq 1$).

The development of the two conflicting activities makes sense provided that the two regions have a larger total income

$$P_{ui} + P_{di} - (D_d + C_u) > P_{uo} + P_{do} \qquad (2)$$

and each region can accept cost sharing in the context of the controversy over pollution, provided that its own income is higher.

$$\begin{cases} P_{ui} - \ell\,(D_d + C_u) > P_{uo} & (3) \\ P_{di} - (1-\ell)\,(D_d + C_u) > P_{do} & (4) \end{cases}$$

or $\begin{cases} P_{ui} - mC_u > P_{uo} & \text{(5)} \\ P_{di} - (1-m) C_u - D_d > P_{do} & \text{(6)} \end{cases}$

Then

$$\ell = \frac{P_{ui} - g(P_{di} - D_d - C_u)}{(1+g)(D_d + C_u)} = \frac{P_{ui} - gP_{di}}{(1+g)(D_d + C_u)} + \frac{g}{1+g} \tag{7}$$

and

$$m = \frac{P_{ui} - g(P_{di} - C_u - D_d)}{(1+g)C_u} + \frac{P_{ui} - gP_{di}}{(1+g)C_u} + \frac{g}{1+g} \frac{C_u + D_d}{C_u} \tag{8}$$

If $g = 1$, ℓ and m are positive whenever

$$P_{ui} > P_{di} - D_d - C_u \tag{9}$$

In other words, the upstream region should pay all or part of the costs associated with pollution if its income is higher than that of the downstream region when the latter pays all the costs.

In particular if the net income of the downstream region was negative when it had to pay all the pollution costs, the upstream region should share the costs

$$\ell > \frac{D_d + C_u - P_{di}}{D_d + C_u} > 0 \tag{10}$$

or

$$m > \frac{D_d + C_u - P_{di}}{C_u} > 0 \tag{11}$$

This rule corresponds to the general principle that water usage in the upstream country should not be such as to ruin the economy of the downstream country.

If the income of the downstream region is higher than that of the upstream region,

$$P_{di} - P_{ui} > D_d + C_u \tag{12}$$

the upstream region would not have to bear any liability for costs associated with pollution ($\ell = m = 0$).

The increases in income of the two regions are equal if

$$P_{ui} - P_{uo} - \ell(D_d + C_u) = P_{di} - P_{do} - (1 - \ell)(D_d + C_u) \tag{13}$$

or

$$P_{ui} - P_{uo} - mC_u = P_{di} - P_{do} - (1-m)C_u - D_d . \tag{14}$$

Then

$$\ell = \frac{(P_{ui} - P_{uo}) - (P_{di} - P_{do})}{2(D_d + C_u)} + 1/2$$

128

and

$$m = \frac{(P_{ui}-P_{uo})-(P_{di}-P_{do})}{2C_u} + \frac{D_d+C_u}{2C_u}$$

The upstream country should pay part of the costs associated with pollution (l > 0 or m > 0)

if $\quad P_{ui}-P_{uo} > P_{di}-P_{do}-(D_d+C_u)$ \hfill (15)

i. e. when its additional income from pollution-generating activities in the upstream region is greater than the additional income of the downstream region after the latter has paid all the pollution costs.

Remarks

1) These formulae for cost sharing have some appeal in that they lead to uniform improvement of welfare. However, two countries may not agree to these formulae when they each seek to improve their own welfare and in particular, when one country does not wish to remain less developed than another country.

2) These formulae imply that developed countries have to pay, to the less developed countries, all the pollution damage costs connected with the development of the developed country.

3) Instead of considering the income of a region, one could compare the per capita income in each region.

4) The main weakness of this system is that it links pollution cost sharing with general problems of assessing the economic situations of neighbouring regions at two points of time where the first is not even clearly defined.

Application to the specific case described in Section 10 (g=1)

In Case A, $P_{di} > P_{do} = P_{uo}$ and $D_d+C_u = 0$ when industrialization starts. Thus, l and m are initially equal to zero. The upstream region may start to industrialize without giving consideration to the damage caused in the downstream region. As industrialization proceeds, the difference between the income of the two regions will decrease and possibly reach D_d+C_u if pollution damage becomes important:

$$P_{di}-P_{uo} > P_{di}-P_{ui} \longrightarrow D_d+C_u$$

The income of the downstream region will always be smaller that it was before the upstream region was industrialized and this might cause some resentment. If industrialization in the upstream region continues to increase, the upstream region should relieve the downstream region of part of the pollution costs (l and m positive).

In Case B, $P_{ui} > P_{uo} = P_{do}$ and $D_d+C_u = O$ when region D decides to change its activities. If region D wishes to develop tourism (Case Bb), l and m are initially positive because the upstream region has a higher income. The upstream region should bear all the pollution costs up to the point where the difference between the incomes of the two regions becomes sufficiently small:

$$\ell = 1 \text{ if } P_{ui}-P_{di} = D_d + C_u$$

$$m = 1 \text{ if } P_{ui}-P_{di} = C_u - Dd .$$

When tourism is further developed, the costs should be divided between the two regions:

i. e. $\qquad 0 < \ell < 1$

if $\qquad -(D_d + C_u) < P_{ui}-P_{di} < D_d + C_u$

and $\qquad 0 < m < 1$

if $\qquad -(D_d + C_u) < P_{ui}-P_{di} < C_u - D_d$

If region D wishes to develop industry (Case Ba), $C_u + D_d = 0$ and there is no problem of transfrontier pollution.

This example shows that the cost sharing formula depends on whether the upstream region started its pollution-generating activities before or after the downstream region developed its pollution-sensitive activities. In the first instance the upstream country may initially develop its polluting industry without bearing any pollution costs, and in the second instance it should initially bear the full cost of abating the pollution from its existing industry.

2. Equitable sharing of pollution costs

Equity would seem to require that the regions should share pollution costs equally or in proportion to the benefits derived from the introduction of activities generating a transfrontier pollution issue. Such a formula would be useful provided that it is a simple one and is widely applicable under the basic assumptions

$$P_{ui} \geqq P_{uo} \tag{16}$$

$$P_{di} \geqq P_{do} \tag{17}$$

$$P_{ui} + P_{di} - T > P_{uo} + P_{do} \tag{18}$$

where $\qquad T = C_u + D_d \tag{19}$

2(a) The incomes of the two regions are equally reduced below their maximum values ("equal regret") if

$$\ell (D_d + C_u) = (1 - \ell)(D_d + C_u) \tag{20}$$

or

$$mC_u = (1-m)C_u + D_d \tag{21}$$

Then $\qquad \ell = 1/2 \ (1/2 \text{ CLP principle}) \tag{22}$

$$m = \frac{C_u + D_d}{2C_u} \tag{23}$$

130

If $D_d \ll C_u$, $m \approx 1/2$ (1/2 PPP principle)

This principle of "equal regret" always gives a positive value for ℓ and m, and ℓ is equal to $1/2$. It has the merit of not making direct reference to the income derived from the activities undertaken upstream or downstream. However, this method cannot be used if one country has to bear pollution costs in excess of the difference between incomes $(P_{ui}-P_{uo}$ or $P_{di}-P_{do})$ and it would be considered inequitable when the costs borne by one region would be high in relation to the added income $(P_{ui}-P_{uo}$ or $P_{di}-P_{do})$.

2(b)   The incomes of the respective regions differ from their maximum values by the same percentage ("equal relative regret") if

$$\frac{\ell(D_d+C_u)}{P_{ui}} = \frac{(1-\ell)(D_d+C_u)}{P_{di}} \tag{24}$$

or

$$\frac{m C_u}{P_{ui}} = \frac{(1-m)C_u+D_d}{P_{di}} \tag{25}$$

Then

$$\ell = \frac{P_{ui}}{P_{ui}+P_{di}} < 1 \tag{26}$$

$$m = \frac{C_u+D_d}{C_u} \frac{P_{ui}}{P_{di}+P_{ui}} \tag{27}$$

This principle of "equal relative regret" always gives a positive value for ℓ and m. It has the merit of not making reference to the initial values P_{uo}, P_{do} which are often uncertain. However, this principle will not be considered equitable if one of the two regions has to bear costs in excess of the increase of income.

2(c)   The increases in income of the two regions are proportional to what they would have been if there was no pollution if

$$\frac{P_{ui}-P_{uo}-\ell(D_d+C_u)}{P_{di}-P_{do}-(1-\ell)(D_d+C_u)} = \frac{P_{ui}-P_{uo}}{P_{di}-P_{do}} \tag{28}$$

or

$$\frac{P_{ui}-P_{uo}-mC_u}{P_{di}-P_{do}-(1-m)C_u-D_d} = \frac{P_{ui}-P_{uo}}{P_{di}-P_{do}} \tag{29}$$

Then

$$\ell = \frac{P_{ui}-P_{uo}}{P_{ui}-P_{uo}+P_{di}-P_{do}} < 1 \tag{30}$$

and

$$m = \frac{C_u+D_d}{C_u} \frac{P_{ui}-P_{uo}}{P_{di}-P_{do}+P_{ui}-P_{uo}} \tag{31}$$

131

In this case, ℓ and m are always positive, i. e., each country should bear a fraction of the pollution costs. Quantity ℓ is independent of the damage cost or the costs of the pollution abatement measures decided upon. This formula for cost sharing is generally applicable. The main difficulty lies in evaluating P_{ui}, P_{uo}, P_{di}, P_{do}.

If the two regions are initially in the same economic situation ($P_{uo} = P_{do}$), and if one region develops activities which provide a higher income than those developed in the other region, the first region will have to bear a larger fraction of the pollution costs. However, its net income will remain higher (thus maintaining an incentive to develop activities that produce a high income).

If one region does not modify its activities, all pollution costs should be borne by the region which does change its activities (whether it be the pollution-generating region or the polluted region). In the case of a change in the activities in the downstream region, this formula implies the recognition of the "Victim Pays Principle" ($\ell = 0$) and in the case of a change in the upstream region, it implies recognition of the "Civil Liability Principle" ($\ell = 1$). This formula would correspond to the "Polluter Pays Principle" if

$$\frac{P_{ui} - P_{uo}}{P_{di} - P_{do}} = \frac{C_u}{D_d} \, ,$$

a condition which would rarely be satisfied.

3.   Determination of a cost sharing system which favours optimum economic development of a boundary region

In this section we consider the following four situations, numbered (1), (2), (3) and (4):

1)   each region has activities producing incomes P_{uo}, P_{do} and the upstream country does not discharge pollutants;

2)   the upstream region replaces its traditional activities by pollution-generating activities producing an income $P_{ui} > P_{uo}$; the downstream region does not modify its activities and is not affected by pollution;

3)   the downstream region replaces its traditional activities by activities which require a clean environment and produce an income $P_{di} > P_{do}$; the upstream region does not modify its activities and thus does not discharge any pollutant;

4)   the upstream region replaces its traditional activities by pollution-generating activities producing an income $P_{ui} > P_{uo}$; the downstream region replaces its traditional activities by activities which require a clean environment and in this case produce an income $P_{di} > P_{do}$; pollution crosses the frontier dividing the two regions and causes a reduction of income T when the level of pollution is optimum.

Let us define the optimal situation as being the alternative which produces maximum income to the two regions considered together.

Table 1 lists the optimal situations according to the additional income accruing to each region as a result of the change in its activities.

Table 1.   OPTIMAL SITUATION (2), (3) or (4)

		$P_{ui}-P_{uo}-T$	
		-	+
$P_{di}-P_{do}-T$	-	$P_{ui}-P_{uo}>P_{di}-P_{do}$ (2)	(2)
		$P_{ui}-P_{uo}<P_{di}-P_{do}$ (3)	
	+	(3)	(4)

3(a)   System based on acquired rights

This system is based on four rules:

a)   the party making a move which creates a pollution problem is responsible for all pollution costs(T);

b)   the party making a move which eliminates a pollution problem will receive a payment from the other party equal to the pollution costs avoided;

c)   if each party simultaneously makes a move which separately would create a pollution problem, each party will be responsible for half of the pollution costs (T);

d)   no payment will be made to the party which moves back to the initial situation if such a move does not eliminate a pollution problem.

Rule (a) is a recognition of the concept of acquired rights. If the move consists in establishing a pollution-generating industry, the other country will be compensated (Civil Liability Principle) because it has an acquired right to a clean environment. If the move consists in creating a pollution-sensitive activity, the other country will not be liable for pollution costs because it has an acquired right to pollute the environment (Victim Pays Principle). As downstream countries cannot hope to always receive crystal-clear waters, such a principle may be considered acceptable.

133

Rule (b) is the reciprocal of Rule (a) and it sets up a mechanism which may be regarded as an automatic "bribing" mechanism. Rule (c) is needed to resolve the difficulty arising when two moves under Rule (a) occur simultaneously. Rule (d) is aimed at keeping the same income in situation (1) irrespective of the strategies adopted to arrive at this situation.

The various economic situations of two countries using the above rules are given in Table 2. In each situation, the sum of the incomes of each country is equal to the total income if the two countries are considered together.

Table 2.   INCOME IN THE VARIOUS SITUATIONS

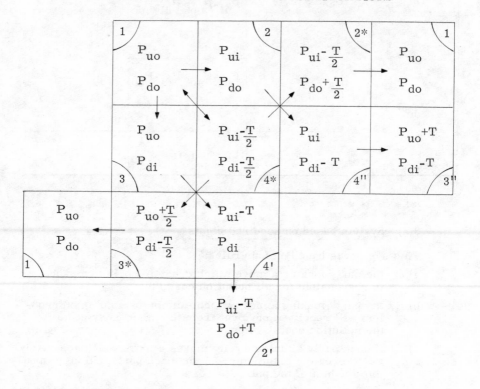

Explanation:

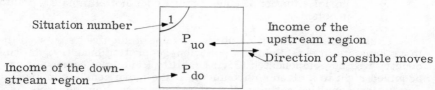

Situation number

Income of the upstream region

Direction of possible moves

Income of the downstream region

Situations 4' and 4" are different because they correspond to historically different development strategies (acquired rights by the upstream or

134

downstream countries). The same holds true for groups 3, 3", 3* and 2, 2', 2*. Situation 4* is the result of a double move which can be followed by situation 2* or 3*.

Having defined the possible situations, let us now examine the sequences of moves which will occur according to the above rules under the following conditions:

a) the first move from situation 1 is arbitrary (but will take place because situation 1 is the least satisfactory situation) and implies no penalty;

b) a move from situation 1 to situation 4* is possible if the two countries act simultaneously under condition (a) above;

c) subsequent moves will be made by the country which increases its income by changing its activity;

d) a move from situation 4* to situation 1 is possible if the two countries act simultaneously under condition (c) above;

e) no move will take place when condition (c) above is not satisfied.

The various strategies depend on the relative values of the income of each country. Table 3 gives the list of such moves. It is seen that whenever $P_{di}-P_{do}$ or $P_{ui}-P_{uo}$ is greater than T, the two countries end up in the optimal situation. If $P_{di}-P_{do} < T$ and $P_{ui}-P_{uo} < T$, there might be cases where the countries do not choose the optimal situation (because the first move was not the best move). This seems unavoidable if the first move is arbitrary i.e., if a country does not need to negotiate with the other country on developments which have no immediate effect on the other party.

Table 3.　DEVELOPMENT STRATEGIES

$P_{di}-P_{do}$		$P_{ui} - P_{uo}$		
		$(< \frac{T}{2})$	$\frac{T}{2}$	T　$(> T)$
$(> \frac{T}{2})$ / $\frac{T}{2}$		1-2 1-3 1-4*-3*-1 ... 1-4*-2*-1 ... 1-4*-1	1-2 1-3 1-4*-3*-1 ... 1-4*-2* 1-4*-1 ...	1-2 1-3-4'-2' 1-4*-2*
$\frac{T}{2}$ / T		1-2 1-3 1-4*-3* 1-4*-2*-1 ... 1-4*-1 ...	1-2 1-3 1-4*-3* 1-4*-2* 1-4*-1 ...	
T $(< T)$		1-2-4"-3" 1-3 1-4*-3*		1-2-4" 1-3-4' 1-4*

Explanation: 1-2-4' means that the regions move from situation 1 to situation 2 and then to situation 4' (as described in Table 2).

135

3(b) System based on cost sharing

The system based on cost sharing uses the rule that the optimal pollution costs are shared between the upstream polluting country and the downstream polluted country according to a fixed formula (e. g. , "Polluter Pays Principle", "Civil Liability Principle" or "Victim Pays Principle"). In this case the upstream country has a net income $P_{ui}-x$ and the downstream country $P_{di}-y$ with $x + y = T$ when they are in situation 4. This system makes no reference to anteriority or acquired rights and has only 4 types of income distribution (Table 4). Moves will take place as in the system described above. Table 5 lists all possible moves. Comparing Table 5 with the end move in Table 3, it is seen that:

a) this system provides income distribution independent of the strategy;

b) it authorizes moves from the optimal situation to a suboptimal situation;

Example: 1-2-4 when 2 is the optimal situation

$$(P_{di}-P_{do} > y, \ P_{ui}-P_{uo} > T); \quad \text{and}$$

c) it does not always lead to the optimal situation

Example: 1-3-4 when 2 is the optimal situation

$$(P_{di}-P_{do} > y, \ P_{ui}-P_{uo} > T).$$

Table 4. INCOME IN THE VARIOUS SITUATIONS

		Upstream	
		Traditional farming	Industrialization
Downstream	Traditional farming	1 P_{uo} P_{do}	2 P_{ui} P_{do}
	Irrigation	3 P_{uo} P_{di}	4 $P_{ui}-x$ $P_{di}-y$

$x + y = T$

Table 5.   DEVELOPMENT STRATEGIES

$P_{di}-P_{do}$		$P_{ui}-P_{uo}$		
		< x 　x	> x 　T	> T
< y		1 3 1 2 1 4 3 1 4 2 1 4 1 ...	1 3 4 2 1 2 1 4 2	1 3 4 2 1 2 1 4 2
> y 　y 　T		1 3 1 2 4 3 1 4 3	1 3 4 1 2 4 1 4	1 3 4 1 2 4 1 4
> T		1 3 1 2 4 3 1 4 3	1 3 4 1 2 4 1 4	1 3 4 1 2 4 1 4

Explanation: 1-3-4 means that the regions move from situation 1 to
situation 3 and then to situation 4 (as described in Table 4).

The deficiency of this system with regard to the previous system
will not appear if:

$$\begin{cases} P_{ui}-P_{uo} > T \\ y = T \end{cases}$$

or

$$\begin{cases} P_{di}-P_{do} > T \\ x = T \end{cases}$$

or

$$\begin{cases} x = T \\ y = T \end{cases}$$

The last possibility should be ruled out because it was assumed
that $x + y = T$. The first two possibilities correspond to the development
of activities producing high income. As this assumption cannot be made
in general, the present system may appear inferior to the first system
although it has the merit of providing an income distribution independently of the strategy. An improvement would result if one region could
"bribe" the other but this may be fairly difficult to implement on an international scale.

137

If conditions (16) (17) (18) are satisfied (i.e., situation 4 is preferable to situation 1), and if x and y are given by:

$$x = \frac{P_{ui}-P_{uo}}{P_{ui}-P_{uo}+P_{di}-P_{do}} \quad T$$

$$y = \frac{P_{di}-P_{do}}{P_{ui}-P_{uo}+P_{di}-P_{do}} \quad T$$

it is seen that $P_{ui}-P_{uo} > x$ and $P_{di}-P_{do} > y$. Hence, the two countries will always choose to develop activities and to create a transfrontier pollution problem. Such behaviour is rational only if $P_{ui}-P_{uo} > T$ and $P_{di}-P_{do} > T$. In the other cases, the present system creates an environment problem without the excuse of increasing the overall income. It might be better to have either a pollution-generating industry upstream or a highly touristic region downstream rather than both at the same time.

In conclusion, due consideration should be given to the first system described above because it is based on traditional legal principles and leads to economically optimal development of neighbouring regions. In such a system, neither the upstream country nor the downstream country is considered "guilty" and neither country has to bear obligations linked to its geographical position. As opposite principles of cost sharing are used, no recognition is given to any of the standard principles. Therefore such a system would appear to offer a compromise between the claims of upstream and downstream countries.

Annex IV

A MULTILATERAL SYSTEM FOR COST SHARING
BASED ON THE PRINCIPLE OF RECIPROCITY

Under the principle of reciprocity, two countries may agree each to undertake a project of the same cost and of interest solely to the other country. For instance, a downstream country may agree to build a road to the upstream country if the upstream country would reduce pollution of the river crossing the frontier. When there is a difference in the cost of the two projects, the country asking for the more costly project should offer some financial compensation. This principle of reciprocity has often been used for solving cases of transfrontier pollution but it implies that the two countries must find a pair of projects of similar magnitude (in cost and in benefit). Until such a pair of projects is found, no agreement can be reached and the economic welfare of the two countries considered together would be smaller.

To overcome this difficulty, it would be useful to introduce a system whereby each country would undertake projects provided that the other countries would reciprocate on a collective basis. This system cannot be applied to countries which are "upstream" as regards all the pollutants and are independent of all their neighbours (autarcic economy on a mountain) because such countries will not find any projects of interest to themselves and they will only accept to undertake projects on behalf of other countries for which they receive payment. This exceptional case should not prevent examination of the proposed system since most countries are politically and economically interdependent.

The proposed system is applicable to situations where all projects under consideration imply a benefit to one country and a cost to another country. The "rules" of the system are:

a) a country requesting a project of cost C_1 to be undertaken in another country will pay to the other country $(1 - \beta)C_1$ and deposit βC_1 in a "bank";

b) the other country will borrow from the "bank" βC_1 and undertake the project (Cost C_1);

c) the deposit by the requesting country will be given a negative interest. The value of the deposit at time t is $C_1 e^{-\gamma t}$ and the liability of the other country to the bank is $C_1 e^{-\gamma t}$ at time t;

139

d) any country which has borrowed more from the bank than it has deposited may propose a new project (Cost C_2);

e) if the country where the project is located agrees to this proposal, the requesting country will pay $(1 - \beta)C_2$ to the other country and will deposit βC_2 in the bank. The other country will borrow βC_2 from the bank and its liability to the bank will be made equal to what it would have been if the two projects were undertaken at the time the second project was agreed upon;

f) if the other country refuses without valid reason to undertake the second project, it loses its deposit in the bank. The country where the first project was carried out is entitled to terminate its operation and has no liability towards the bank in respect of this project.

Notes

1) Similar rules can be devised without introducing a negative interest rate when the rate of inflation is higher than the rate of interest.

2) Estimates of the benefit are not needed explicitly. Each country will presumably arrange the projects in the order of descending benefit/cost ratio. Neighbouring countries with economic conditions that are usually similar will undertake projects that have similar benefit/cost ratios.

Examples

a) Let us assume that country 1 asks country 2 to undertake a project of total cost C_1. Initially the cost to country 1 is $(1 - \beta)C_1$ and ultimately the cost is $(1 - \beta)C_1 + \beta C_1(1 - e^{-rt}) \longrightarrow C_1$. Country 2's initial cost is C_1 and ultimately it is $C_1 e^{-rt} \longrightarrow 0$.

b) If a pair of projects of cost C_1, C_2 is introduced simultaneously, the cost for country 1 at time t is $C_1 + \beta e^{-rt}(C_2 - C_1)$ a quantity which varies between $(1 - \beta)C_1 + \beta C_2$ and C_1. Similarly the cost for country 2 at time t is $C_2 + e^{-rt}(C_1 - C_2) \longrightarrow C_2$.

If country 1 is polluted by country 2, and if country 1 wishes country 2 to be financially responsible for all pollution control measures, country 1 will ask that $\beta = 1$. Thus country 1 has obtained temporary recognition of the Polluter Pays Principle. Country 2 could accept this formula since in the end it is country 1 that will pay the project cost and country 2 will not have to provide any additional finance for undertaking this project.

Country 2 may not agree to bear any responsibility at any time and could ask that $\beta = 0$. Country 1 would thus have to bear immediately

140

the full cost of pollution abatement measures and to accept the victim pays principle.

Between these two extreme positions, the value $\beta = 1/2$ could be chosen. The two countries would then have initially the same liability (country 1 pays $\dfrac{C_1}{2}$ or $\dfrac{C_1 + C_2}{2}$ and country 2 owes $\dfrac{C_1}{2}$ or $\dfrac{C_1 + C_2}{2}$ according to the number of projects undertaken).

If $C_1 = C_2$, the costs to each country are the same for any value of β. It is thus not necessary to state which principle the countries have adopted.

c) If three countries lie along the same river and if there is a project of cost C_1 to be undertaken in country 2 for the benefit of country 1, and a project of cost C_2 to be undertaken by country 3 for the benefit of country 2, the costs are:

Country 1:

$$C_1 + \beta e^{-\gamma t} C_1 \longrightarrow C_1 \tag{1}$$

Country 2:

$$C_2 + \beta (C_1 - C_2) e^{-t} \longrightarrow C_2 \tag{2}$$

Country 3:

$$\beta e^{-\gamma t} C_2 \longrightarrow 0 \tag{3}$$

When $C_1 = C_2 = C$, country 2 pays exactly the same amount as that it would have to pay whether the "Polluter Pays Principle" or the "victim pays principle" is adopted. For country 2, the same system should be adopted with regard to the upstream country 3 and the downstream country 1 in order not to pay an amount different from C.

d) If in addition country 3 asks country 1 to undertake a project of benefit solely to country 3, the costs are:

Country 1:

$$C_1 + \beta e^{-\gamma t} (C_3 - C_1) \tag{4}$$

Country 2:

$$C_2 + \beta e^{-\gamma t} (C_1 - C_2) \tag{5}$$

Country 3:

$$C_3 + \beta e^{-\gamma t} (C_2 - C_3) \tag{6}$$

If $C_1 = C_2 = C_3 = C$, and for any given value of β, the three countries would agree to undertake the three projects and would be better off. The proposed system enables them to find a multilateral solution when no bilateral solution is possible.

e)   If country 1 wishes that projects of cost C_{1i} be undertaken by country i for the sole benefit of country 1, and if country j wishes that a project of cost C_{j1} be undertaken by country 1 for the sole benefit of country j, the cost of all the project to country 1 is

$$\sum_i C_{1i} + \beta e^{-\nu t} (\sum_j C_{j1} - \sum_i C_{1i}), \tag{7}$$

a quantity which varies from $(1-\beta) \sum_i C_{1i} + \beta \sum C_{j1}$ to $\sum C_{1i}$.

As country 1 can propose a project only when $\sum C_{j1} - \sum C_{1i} > 0$, the second term will not be very large and it will fluctuate around zero (when it is negative a new project can be introduced by other countries to make it positive, and when it is positive country 1 can introduce a new request). The ratio of the second term to the first term ($\sum C_{1i}$) will decrease when the number of projects increases. Therefore the system will lead to the situation in which all countries will undertake projects to an amount approximately equal to the amount undertaken for them. If a difference remains the requesting country would have to pay this amount in the end.

f)   If country 1 asks country 2 to undertake a project of cost C_1 and T years later country 2 asks country 1 to undertake a project of cost C_2, the cost to country 1 just before it agrees to the second project is $(1-\beta)C_1 + \beta C_1 (1-e^{-\nu T})$ and the cost to country 2 is $\beta C_1 e^{-\nu T}$. (See figure.)

Country 2 transfers $(1-\beta)C_2$ to country 1 and deposits βC_2 in the bank. In addition, country 2's liability to the bank is increased by the amount $\beta C_1 (1-e^{-\nu T})$ and becomes βC_1. Simultaneously the deposit account of country 1 is increased to βC_1 by adding $\beta C_1 (1-e^{-\nu T})$. Country 1 does not have to finance the second project.

At time $t = T + \Delta t$, the cost to country 1 is

$$C_1 + \beta (C_2 - C_1) e^{-\nu \Delta t} \longrightarrow C_1 \tag{8}$$

and the cost to country 2 is

$$C_2 + \beta (C_1 - C_2) e^{-\nu \Delta t} \longrightarrow C_2 \tag{9}$$

If country 1 refuses to undertake the project requested by country 2, it loses its deposit $\beta C_1 e^{-\nu T}$. If $C_2 < C_1$, it is therefore in the interest of country 1 to accept the project. If $C_2 = C_1$, the cost to country 1 would be the same. If $C_2 > C_1$, country 1 could refuse the project

142

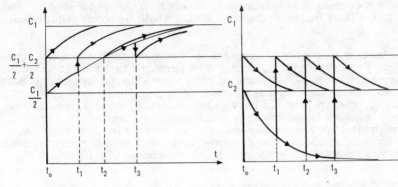

Cost to country 1 as a function of time when country 2 introduces a project of cost C_2 at time t_0, t_1, t_2 or t_3. $\beta = \frac{1}{2}$ $C_2 = \frac{1}{2} C_1$

Cost to country 2.

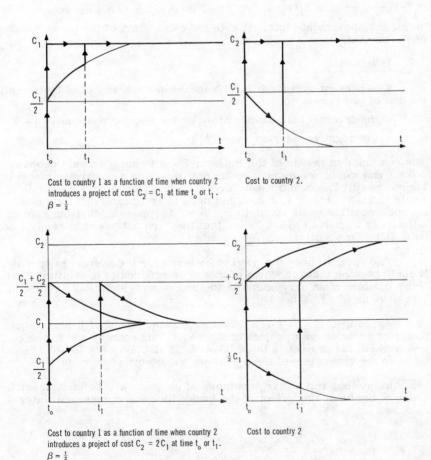

Cost to country 1 as a function of time when country 2 introduces a project of cost $C_2 = C_1$ at time t_0 or t_1. $\beta = \frac{1}{2}$

Cost to country 2.

Cost to country 1 as a function of time when country 2 introduces a project of cost $C_2 = 2C_1$ at time t_0 or t_1. $\beta = \frac{1}{2}$

Cost to country 2

because it corresponds to a larger project than that requested by country 1 in the first instance. Such a refusal would be permissible when the ratio C_2/C_1 is too high.

Another tactic for country 1 is to delay the implementation of the project. To overcome this, it might be arranged that the account of country 1 is readjusted at some point of time after the project proposal is put forward rather than at the time of implementing the project. This penalty for country 1 would have to be decided by some agreed arbitration court. If country 1 introduces administrative difficulties, the cost of the second project may rise substantially and become too high for country 1. Here again, there should be recourse to a court in order to determine whether the cost of the project is not made abnormally high.

g)   Initial cost and operating cost

A project relative to transfrontier pollution usually implies an initial investment and sometimes annual expenditures which do not need to be combined in a single figure. The requesting country would thus ask the other country to undertake a project implying costs $C_1(0)$, $C_1(1)$, $C_1(n)$ during years 0, 1, n. With this method, it is not necessary to define the appropriate interest rate for calculation of the present value of the project.

Discussion

The present system differs from those generally used by the introduction of two parameters:

- the fraction paid immediately by the requesting country $1 - \beta$
- the negative interest rate "γ".

The countries in favour of the Polluter Pays Principle would choose $\beta = 1$ and would see no reason for not receiving a positive interest on their deposit. The countries in favour of the "victim pays principle" would choose $\beta = 0$ (γ does not need to be specified). They can accept a positive value of β ($0 < \beta < 1$) provided that the rate of interest is negative ($\gamma > 0$) and that they are not asked to repay loans until some date sufficiently far ahead.

The rate γ does not need to be large for the system to operate. It could be such that the worth of money is reduced to 0.01 after 50 years. If the lifetime of an agreement on the proposed system is 25 years, a larger value of γ is needed.

Agreement on $\beta = 1/2$ would have some appeal if it gives equal power of decision on a project to the requesting country and to the country in which the project is to be located. It also amounts to a compromise between the viewpoints of the upstream and downstream countries.

In the long run, a large number of projects will be undertaken because of the balancing effect of the bank which authorizes multilateral solutions

Generalizations regarding projects of common interest

A project of common interest of cost C brings a benefit B_1 to country 1 and a benefit B_2 to country 2. If country 1 and country 2 are undertaking projects for which the cost/benefit ratio is $a_1 \leq 1$ and $a_2 \leq 1$, country 1 is prepared to pay no more than $a_1 B_1$ and country 2 is prepared to pay no more than $a_2 B_2$. If $a_1 B_1 + a_2 B_2 \geq C$, the two countries can agree to undertake the project. A cost sharing formula would be that country 1 should pay $\dfrac{a_1 B_1}{a_1 B_1 + a_2 B_2} C = C_1$ and country 2 should pay $\dfrac{a_2 B_2}{a_1 B_1 + a_2 B_2} C = C_2$

If $b = \dfrac{a_1 B_1}{a_2 B_2}$ and $b < 1$, the project of common interest of cost C can be regarded as a pair of projects of cost $\dfrac{b}{1+b} C$ requested by each country and an additional project of cost $\dfrac{1-b}{1+b} C$ requested by country 2 of country 1. Therefore the system described above can be used for projects of common interest. However, the countries may disagree on the relative cost sharing because this depends on an estimate of the benefits in both countries and the fact that the two countries use different benefit/cost ratios.

Projects of common interest do not need to be disaggregated into a pair of equal projects and an additional project when the cost agreed by the countries C_1 and C_2 is also the cost of parts of the projects in each country $\overline{C}_1$, $\overline{C}_2$. If this is not the case, it may be convenient to subdivide the common project into a number of pairs of subprojects of equal cost, plus an additional subproject.

Examples

If $C_2 > C_1 > \overline{C}_2$ and $\overline{C}_1 > \overline{C}_2$, the two countries are carrying out two equal subprojects of cost $\overline{C}_2$ in each country and two equal subprojects of cost $C_1 - \overline{C}_2$ in country 1. In addition, country 2 requests that an additional subproject of cost $C_2 - C_1$ be undertaken in country 1.

If $\overline{C}_1 < C_1 < C_2 < \overline{C}_2$, the two countries are undertaking two equal subprojects of cost $\overline{C}_1$ in each country and two equal subprojects of cost $(C_1 - \overline{C}_1)$ in country 2. In addition, country 2 undertakes a subproject of cost $C_2 - C_1$ on its territory and for its own benefit.

CONCLUSION

The proposed system makes it easier to arrive at multilateral solutions to problems of one-way and two-way externalities by incorporating them into a general system based on reciprocity. It can operate even when the pollution-generating country does not recognize responsibility for damage caused to other countries and it leads to an improvement in the general welfare. It should therefore be acceptable to all countries, whether upstream or downstream.

Downstream countries could, however, hope to have a more favourable system if some international rule were adopted whereby upstream polluters would have to bear all the pollution costs. As long as such a rule is not adopted, the system described here represents an improvement on the present situation.

OBSERVATIONS ON THE ECONOMICS
OF TRANSNATIONAL ENVIRONMENTAL EXTERNALITIES

by

Ralph C. d'Arge

University of California, Riverside,
and Resources for the Future, Inc., U.S.A.

I[1]

As man socialized and the human population clustered into geographically, culturally, and politically defined nation states, reaping
the material benefits of specialization and efficiency, human interaction
increased both within and among these states. However, because of the
sheer magnitude of the environment and its assimilative capacity in
relation to human populations and industrial production, in the historical
past there was little, if any, interaction between states that involved a
purely non-human or environmental element. A nation's waterborne
wastes were easily assimilated by river, estuary, and coastal waters
before such wastes could interfere with other nations' activities. No
state produced and consumed chemical compounds that were not easily
neutralized by a seemingly infinite natural environment. In the past few
decades, such a perception of the natural environment has increasingly
changed to the realization that the planet's environmental assimilative
capacity, even augmented by substantial investment, cannot sustain an
endlessly growing population or material consumption per capita. Nations
are coming to realize that not only are they economically and politically
interdependent but also environmentally dependent with no well defined
international markets or political mechanisms for efficient regulation.

Economic science has analyzed problems of non-market interdependence for a very long time, applying the concept of externalities
since Pigou. In essence, externalities are social interdependencies not
taken into account by formal markets or by agreements between the affected individuals or nations. Thus, externalities embrace a spectrum
of problems from the neighbour's noisy and disturbing stereo to the commitment of all nations today toward rapid economic development which
may preclude choices between material consumption and aesthetic enjoyment of future nations. In either case, the affected party's, i.e., the
disturbed neighbour's or future generations', preferences are not adequately being considered when a decision is made.

The classical economic solution to such externality problems was
to "internalize" them by either developing a well-defined market for the
"spill-overs" or controlling them through collective provision of regulations. Neither of these possibilities appears easily amenable to the
problem of transfrontier externalities in general, and environmental
externalities in particular. First, environmental externalities have
arisen because most dimensions of the natural environment on a regional
or global scale are resources without rigidly defined or enforceable
ownership rights. The oceans, atmosphere, and electromagnetic spectrum

1. I am indebted to Anthony Scott and Serge Kolm, without implying agreement, for comments
and corrections on an earlier draft.

are typical examples. These resources are viewed as being commonly owned or not owned at all. A nation which agreed to a particular pattern of ownership of these resources could potentially lose some of its implicitly controlled resources and thereby national wealth.[1] As long as international entitlements are obscure, any nation can lay implicit claim to the common property resource exceeding any equitable share it may presume to receive if entitlement were made explicit. This is not to say that once some other nation impinges on a nation's perceived implicit entitlement, that it will not find a negotiated settlement and thereby explicit entitlement to be superior to an implicit one. However, the impinging nation, in negotiating, must revise downward its own perceived ownership of the common property resource. In consequence, proceeding from a situation of implicit entitlements of common property resources to explicit regulation and thereby ownership means that some (or all) nations must revise downward their expectations of national wealth, stemming from the resources that each implicitly believes it controls.

A second aspect of major importance arises from the concept of national sovereignty. Not unlike consumer sovereignty as conceptualized by economists, national sovereignty implies the idea that governments, acting in their own interest will, omitting deviations in power or information implying political or economic monopoly, achieve the greatest welfare for all by independently pursuing autonomous goals and interacting through international markets. The belief in national sovereignty as an ideal is so imbedded that it is impractical to presume it will be easily given up.

Coupling the concepts of national sovereignty in decisions and the idea that implicit, as opposed to explicit, entitlement of international common property resources yields a greater perceived wealth for nations, is suggestive that resolution of transnational externality problems will generally need to be embedded within the following restrictions:

1) No nation will easily accept international agreement on entitlement of significant common property resources without compensating payments to retain its perception of national wealth. In consequence, the classical answer to externality problems of internalizing the decision-making process for the resource is not easily transferable to transnational problems. A new overriding element of distributional gains and losses must be simultaneously included in efficiency considerations.

2) Unidirectional transnational externalities, if they are of substantial importance to the emitter country as a method of waste disposal, will in general be resolved by some form of a compensation system where compensation flows from the receptor country. The non-liability case (or "victims must pay" principle) will generally be dominant.[2]   This is to be contrasted

---

1. Christy draws a very useful distinction between the production of wealth and distribution or ownership of wealth with regard to ocean fisheries. The first concept involves issues of access and free use while the second involves specification of shares. The discussion in this paper will be centered on distributional as opposed to use issues. See F. T. Christy, Jr., "Fisheries: Common Property, Open Access, and the Common Heritage, " Pacem in Maribus (The Royal University of Malta Press, 1971), Ch. 6.

2. The appearance of reciprocity may negate this statement, particularly in those cases where an external diseconomy in one direction between countries is offset by an external economy in the opposite direction.

with the reciprocal environmental externality case where compensation may flow in either or both directions.

3) International court settlements of transnational externalities are not likely to yield satisfactory results. There appear to be three almost insurmountable problems. First, how are damages to be measured and damage payments assessed? The receptor country's social values may be strikingly different than the emitter country's. In consequence, there may not be a social welfare index that is applicable to both. If the damage is entirely confined to hindering production in the receptor country, then international trading prices, at least at the margin, offer a measure of welfare loss. However, if the impact is on individual citizens with no market prices representing their losses, then a measure of welfare loss is not available except through direct examination and questioning. In addition, there may be uncertainty as to the magnitude of loss unless the externality is allowed to continue to the point of maximum damages, i.e., threshold levels of fish populations. Second, given the sovereign rights of nations, no nation can be forced to pay environmental damages. The tradeoff here is in terms of loss of international prestige and goodwill or increasing the possibility of conflict versus monetary payments based on possibly misrepresented public preferences in the receptor nation. Third, there is basically the "chicken and egg" problem of historical precedent most dramatized by airports and noise pollution. An airport is built drawing in people that then are affected by airport noise. As Coase cogently stated the problem, who is responsible? Who is the polluter in the "polluter must pay principle"? As environmental problems increase in severity and potential damages induced between countries rise, it seems that assigning responsibility will become increasingly difficult. In this context, there is also the problem of assigning damages when more than one nation's waste residuals contribute to total damages. If the different nations' residuals are synergistic or if damages are non-linear relationships of waste intensity, then there is no easy method of determining how much responsibility each nation should take even with the "polluter must pay" principle. Thus, it can be anticipated that international courts or international commissions will have difficulty in arbitration even if such institutions were given some regulatory powers.

To conclude this rather negative introduction, transfrontier environmental externalities are not likely to be resolved by international organisations without some form of compensation to both countries - an unlikely case unless a transnational external economy can be discovered to offset the inherent costs of a transnational external diseconomy. In this paper, I attempt to develop a taxonomic discussion of models on transnational environmental externalities and analyze them in the context of bilateral and multilateral negotiations. The role of international tribunals, courts, or management commissions is not analyzed explicitly except as an occasional point of reference for "ideal" efficient utilization and distribution of environmental resources.

Wealth Effects of Transnational Externalities in Production

In order to clarify taxonomically transnational externalities, we shall employ the traditional concepts of production and utility relationships in economics. In so doing, externalities will be divided into four categories:

1. externalities generated in production processes that affect production costs and processes in other nations. Examples include industrial water pollution that requires treatment before industrial use in a downriver nation or the salt pollution of the Colorado River by U.S. agricultural tailings' water that reduces farm productivity in Mexico;

2. externalities generated in production in one country that do not affect production but degrade the environment of citizens in other countries: an example is the acid rains in Scandinavia emanating from the Ruhr industrial complex;

3. externalities generated by acts of consumption or final use of goods which affect production costs, i. e. , rude tourists;

4. externalities generated by acts of consumption which influence environmental quality in other countries, i. e. , municipal wastes affecting recreational use of Lake Erie. It appears relevant that for transfrontier environmental externalities the first two cases could be considered the dominant ones currently. However, as urban sprawl continues and populations centralize and enlarge, pollution from acts of consumption might become increasingly relevant as a transnational problem. Municipal wastes in the Great Lakes and elsewhere offer support for this allegation.

In addition to classifying externalities on the basis of source such as production processes or acts of consumption, they can also be usefully classified by how each enters utility and production relationships in nations. Definitionally, a truly global externality problem is one in which the externality enters some production function, utility function, or both in each nation. Alternatively, a regional externality problem connotes that the externality only enters a prescribed subset of all nations' utility and/or production functions.

Using the neo-classical concept of utility and production functions, the various types of transnational environmental externalities can be cataloged. Let $F^i(x, k, y)$ denote the production function for country i where x denotes a vector of outputs, k a vector of resource inputs, and y a vector of environmental externalities that influence production but cannot be controlled by autonomous decisions of firms in nation i. The vector y may include such components as water quality in a river, lake, or estuary jointly used by several nations, or a common airshed. Thus, $F^i(x, k, y)$ represents a production frontier for country i, where given its resources k, externality components y which, prior to regulation, it does not control, and levels of production for N-1 of the N products it produces, the function yields the maximum of the Nth product the country can produce. Omitting considerations of environmental externalities within a nation for the moment, i. e. , presuming all y's are externally determined for the nation, then F^i can be viewed as a wealth measure for the ith country. As some or all of the components of y are changed, then so must the perceived and actual wealth of the ith country. In Figure 1, a simple diagram is given where the country produces two goods, and a component of y is changed which influences the production of one of those goods, x_2. Note that y* represents an external diseconomy since maximum production of x_2 is reduced for any predetermined production level of x_1. If both products of country i are affected by the transnational external diseconomy, then the curve in Figure 1 would move more uniformly inward. [1]

A glance at Figure 1 easily confirms several expectations on international trade patterns. Countries affected by uncompensated transnational externalities in production will tend to produce more of those commodities less influenced by external diseconomies if these economies respond to international prices. If a country is relatively small and the transnational externality is also relatively small so that compensation that adjusts trade patterns has no perceptible influence on international trade prices, then a clear measure of loss in wealth of the country can be obtained. If the country produced at point b prior to the transnational externality or in its absence, but with the externality it produces at a, then income loss measured by international prices equals the distance between c and c'. However, if compensation or removal of the externality changes international prices, then the direct linkage between measured income losses and welfare losses can be broken. As an extreme example, let us presume that compensation or payments for resolving the transnational externality in production of commodity x_2 shifts international prices from dd to c'c' in Figure 1. The country thereby shifts its production for g to b. But now it shifts its consumption point from n to h, thus exporting more x_2 for less x_1. The country has been made <u>worse off</u> from the removal of or compensation for the transnational externality because of international price effects. This is obviously an extreme case, but it demonstrates quite simply the problem of determining welfare loss when international prices are affected by the adjustment for external diseconomies. It also underscores the point that even a receptor country may lose by making a commitment to international arbitration of transfrontier

---

1. In the international trade literature, resource endowments traditionally are not presumed to move internationally. Thus, with this assumption transnational environmental externalities will shift the production curve inward but generally not alter its convexity properties regardless of how pervasive the external diseconomy. This can be easily demonstrated. Since an external diseconomy in production may reduce the productivity of inputs, an output shifts from one commodity to another, one can expect diminishing marginal productivities of inputs to continue at an accelerated rate, as production of the second commodity is increased. Consequently, convexity will generally be retained. Recently, Baumol, Starrett, and others have pointed out the highly pervasive externalities in production may destroy the convexity of production curve for commodities if resources <u>are transferable</u> in production. Transnational externalities do not present this problem so long as resources are immobile. This is suggestive that multinational corporations may at some point contribute to non-convexity, making it more difficult to identify efficient intra-country environmental control strategies. See W. J. Baumol and David Bradford, "Detrimental Externalities and Non-Convexity of the Production Set," <u>Economica</u> (May 1972); and D. Starrett, "On a Fundamental Non-Convexity in the Theory of Externalities," Harvard Institute of Economic Research, <u>Discussion Paper 115, 1970</u>.

Figure 1
EFFECT OF TRANSNATIONAL EXTERNALITY
ON A COUNTRY'S PRODUCTION POSSIBILITIES

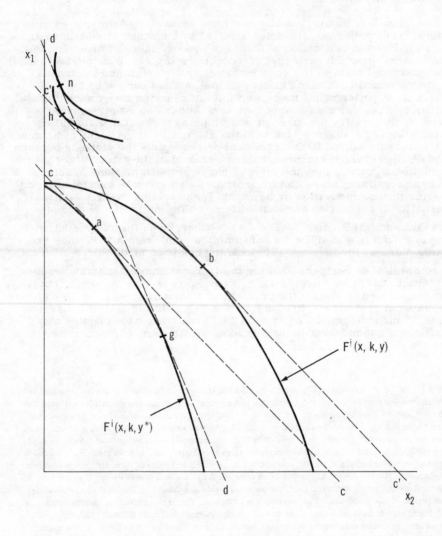

externalities unless international trade effects are given adequate consideration. Normally, however, we should expect that with good information, bilateral negotiation would lead to net welfare gains for both emitter and receptor countries.

A second aspect of consideration regarding wealth effects of transnational externalities in production is how the externality enters production functions in both emitter and receptor nations. Obvious alternatives in the receptor case include:

1) effects on all or some resource inputs, either through reducing their marginal productivity, total productivity, or both;

2) the choice of processes and techniques or factor proportions separable from impacts on the productivity of inputs;

3) adjustment in qualitative aspects of the product which influences its value;

4) the requirement of additional technology or other inputs to alter inputs before they can be used, such as water purification.

Note that 1 and 4 relate explicitly to particular factors of production, while 2 affects the transformation process directly. Finally, under 3, not only must one consider cost-minimizing control strategies but also market demand considerations to measure welfare losses. In terms of the functional relationship F^i discussed earlier, the four impacts can be translated as follows:

Table 1.   IMPACT OF EXTERNAL DISECONOMIES
ON PRODUCTION PROCESSES

IMPACT OF EXTERNALITY	SYMBOLIC REPRESENTATION	COMMENTS
1. Productivity of inputs	$F_{ky^*} < 0$ [1] $F_{xy^*} < 0$	Marginal productivity of inputs decreased Total productivity of all inputs decreased uniformly
2. Process or technique substitution	F replaced by F'	Transformation process of inputs to outputs altered separable of effects on input productivity
3. Change in qualitative aspects of output	$F(x^*, k, y^*)$ with $x^* \neq x$	New vector of outputs x with different qualitative characteristics or dimensions
4. Additional technological adjustment of other inputs	$F(x, k^*, y^*)$ with $k^* = f(x, k, y^*)$	Production function is altered to embody additional processes

1. F_{ky^*} represents the derivative of the production function with respect to input k divided by the derivative of the production function with respect to y^*.

For any particular transnational environmental externality, each impact in isolation or any combination may be operative. However, the policy implications for each in measuring wealth or income losses can be substantially different. For example, in assessing the extent of receptor costs of the externality, if the only impact is a shift downward in productivity of all inputs, then costs are easily measured applying international prices. This is also the case when all that is required is some additional residuals processing. Alternatively, in the case of product charge, costs must be determined as the difference in net rent between production of the qualitatively different commodities, i.e., beer produced in Olympia, Washington, or in St. Louis with variations due to water quality. Such an assessment requires explicit knowledge of consumer preferences in the receptor country and countries it exports to. Likewise, unless there are accurate measures of the cost of process change or change in technique, it will be most difficult to assess receptor nation costs from this source of impact. A simple dynamic example with technological change will suffice to illustrate complexities here. A nation downstream is affected by increasing deterioration in water quality. In consequence, firms there respond by selecting a less water-intensive production process. Embodied technical change in this process leads to greater efficiency in production and efficiencies in production compensate partially or offset completely inefficiencies due to water quality deterioration. Where is the loss to the nation? It surely includes technology conversion costs, but an unanticipated external economy has otherwise offset external diseconomies associated with poorer water quality. This example is cited only to point out the difficulties inherent in measuring the wealth or income loss of the receptor nation. Conceptually, the rule to follow would be the so-called "with and without" principle encountered in benefit-cost analysis. Thus, an appropriate static measuring device of wealth-income loss is output-valued at international prices with and without the transnational externality. However, if there are underlying dynamic processes such as learning embodied in external diseconomies, such static measurements will be disputed in bilateral negotiations particularly with the adoption of the "polluter must pay" principle.

In the emitter country, for the same four cases of impacts on inputs (in a positive sense); the qualitative characteristics of products or shifts in processes are possible because of the country's ability to use environmental assimilative capacity and cause downstream damages without compensation. When the transnational externality is strictly related to production sectors in the several countries, it is significant to ask whether the impact on factor use will be completely symmetrical. In other words, if the marginal productivity of a factor is reduced downstream, is it carried upstream for the same factor or is the upstream country's outward shift in $F^i(x, k, y*)$ induced by different causes? If for example the upstream country by using the river for upstream dumping rather than incurring process changes (while the downstream effect is to reduce the marginal productivity of a single factor downstream), then assessment of damages is relatively straightforward with international prices constant. However, if there are qualitative changes in the upstream country's product (which is competitive or complementary with the downstream country's product that also undergoes qualitative changes), then again demand conditions must be understood to assess damages and measure the involuntary transfer in wealth.

For example, take two countries with equal factor endowments and resource size facing competitive international prices that are constant, i.e., each country is small in relation to global or regional markets. The upstream country increases factor productivity per se by dispersing heat with cooling towers by 10%, while the downstream country experiences the need for cooling towers in order to use river water for production of commodities which requires, by assumption, 8% of its factors of production. Since in neither case by assumption do productivity of individual factors or international prices shift, the composition of output in the two countries would not change, provided certain theoretical conditions on production functions hold. All that occurs is that the emitter country receives a 10% increase in factor income and the receptor an 8% decrease in factor income while the world as a whole undergoes a net 2% rise in the amount of commodities produced by the two countries. If the externality can only be corrected by stopping the heat emission upriver, then it should not be. Allowing the externality to continue is economically efficient but distributionally adverse to the receptor country. Since there generally will be no mechanism to require the emitter to provide compensation, i.e., except by threat of receptor country or altruistic goals of the emitter country, no transfer payment will occur. A "polluter pays" principle similar to the one adopted by OECD Member countries where the polluter must pay for controls and not pay compensation for damages would lead to an inefficient allocation of resources in this case.

What is important to note here is that transnational production externalities which are uncompensated induce distortions of production relationships in both emitter and receptor countries. Perceived wealth of both change as a result of the involuntary character of externalities. However, it is very likely that the impact downstream because of different factor endowments, production technologies, factor mixes, and comparative advantage in commodities will be different than the upstream country's impact. A guess might be that in most cases, the emitter saves in costs of additional technology, i.e., recycling and in-factor productivity per se, while the receptor country tends to lose in terms of required process or technical changes in production and qualitative aspects of the commodities produced. In consequence, the usual implicit assumption that upstream and downstream control costs are similar in magnitude is not likely to be valid. This tentative hypothesis coupled with the observation that some environmental assimilation will break the one-to-one correspondence between emissions and impacts is suggestive that only in particular cases can one assert that cost efficiencies or types of control might be similar.

To summarize this rather scattered discussion on the wealth effects of transnational externalities in production:

1) these types of externalities can be viewed as involuntary transfers of a nation's wealth;

2) if the externality influences competitive international prices, then its resolution by the "polluter pays" principle may economically harm the receptor country as well as the emitter country.

Likewise, the "victim must pay" principle may economically harm both the emitter and receptor countries. However, transnational externalities in production generally yield a rise in income in one country and a fall

157

in income in the other. In certain instances, where control costs or efficiencies are different between countries, it may pay to retain the externality and, if necessary, use lump-sum transfers to achieve equity.

Wealth Effects from Transnational Externalities on Consumers

Thus far, the discussion has been developed from a perspective of externalities in production, but a few words are added here to describe the impact of externalities in final use or consumption as distinct from production. In order to develop the idea of a wealth transfer resulting from transnational externalities in consumption, I apply a simple graphical technique first used by Dolbear.[1] Let there be two countries with collective utility functions as follows:

1) $U^1(x_1, y_2)$

$U^2(x_2, y_2)$

where x_1 and x_2 are the quantities of a private good consumed in countries 1 and 2, and y_2 is the quantity of a private good consumed by 2 but yielding an external diseconomy to 1.

With first and second derivatives denoted by subscript, we assume:

$$U^1_x > 0 \qquad U^1_y < 0$$

$$U^1_{xx} U^1_{yy} > (U^1_{xy})^2$$

$$U^1_{xx} < 0 \qquad U^1_{yy} < 0$$

$$U^2_x > 0 \qquad U^2_y > 0$$

$$U^2_{xx} U^2_{yy} > (U^2_{xy})^2$$

$$U^2_{xx} < 0 \qquad U^2_{yy} < 0$$

Note that the usual strict concavity assumptions are specified for preferences toward the two private goods, x and y. The significant difference, however, is that individuals in country 1 are presumed to undergo increasing marginal disutility if the externality is intensified. For externalities like noise, this may not be the case since tolerance (or environmental adaptation in the Dubos sense) may rise with increasing exposure aided by loss of hearing.

Next, a budget constraint for both countries is postulated:

2) $px_1 = M_1 + B$

$px_2 + ry_2 = M_2 - B$

or

2') $p(x_1 + x_2) + ry_2 = M_1 + M_2$

1. F.T. Dolbear, Jr., "On the Theory of Optimum Externality," American Economic Review (March 1967).

where p and r are the given international prices for private goods x and y, and M_i denotes the initial income of country i. Negotiation for external effects between 1 and 2 is allowed through payment of a bribe B, with $B \gtreqless 0$, which cancels out of equation (2'). For simplicity, we shall assume p = 1 by suitable redefinition of quantity units of x.

In Figure 2, indifference maps are depicted along with initial (before the externality appears) endowments. Country 2's indifference map is identified with the origin 0. However, country 1's indifference map is not. Since 1 cannot exchange y_2 in an international market, its budget constraint would be a sloped line commencing from point a, i.e., the country's exchange price for y_2 is initially zero; and the country will, given a choice, be at a corner solution. As point a moves to the left, country 1 has a higher budget. But in order to represent 1 and 2's combined budget in x-y space, the budget line and indifference map of 1 must be rotated in order to be parallel with the slope of the combined budget line such that the sum of 1 and 2's budget does not exceed their combined budget. Thus, the indifference map for country 1 is rotated in x-y space so that x_1 is always measured from the aggregate budget line from right to left in Figure 2.

Initial consumption is at point a where both 1 and 2 are consuming x but neither is "consuming" y_2. Next, country 2 "purchases" the externality yielding commodity and moves to point b where the budget line aa' is tangent to the indifference curve 22. Clearly, this point is not Pareto optimal in a global sense, since country 1 suffers a loss of utility and is moved involuntarily from indifference curve 1'1' to 11. By negotiation, they could achieve the contract curve cc, which implies a reduction in intensity of the externality. For simplicity, it is assumed that the externality and consumption of y_2 are joint products. It should be reemphasized that country 1 has control only over consumption of product x and cannot either purchase or sell y_2, except by negotiating with 2. Also, and perhaps more importantly, the contract curve defined as cc is not one where marginal rates of substitution for x and y are equated with relative prices defined as 1/r, although the budget line of country 2, adjusted for fees or payments, must pass through this tangency.

What we now wish to introduce are simplified versions of the "third party liability" (TP) and "victim must pay" (VP) principles. First, the TP rule is presumed to require that country 2 compensate country 1 for all disutilities caused by the externality.[1] With negotiation and the TP

---

1. Note that country 2 by assumption can only regulate the externality's impact by reducing consumption of y_2 or paying country 1 some x to tolerate the external diseconomy. Thus, country 2 cannot provide controls and country 1 implicitly cannot undertake defensive expenditures internally to reduce the intensity of the externality. Recently, the OECD Member countries agreed to implement a "pollutor pays" principle which in effect requires the polluting firm to pay all control costs and the receptor firm or consumer to pay all residual damages. Also, transfrontier pollution was explicitly excluded from this agreement. For clarity, we adopt the following terminology: polluter is responsible for all control costs and residual damages is denoted as the "third party principle" where third party connotes the implicit need for an outside enforcement body or court in order to establish exclusive right for the receptor nation; polluter pays only the control costs and no residual damages will be referred to as the "polluter pays" principle. See Organisation for Economic Co-operation and Development, Recommendation of the Council Concerning International Economic Aspects of Environmental Policies, Paris (May 26, 1972).

Figure 2

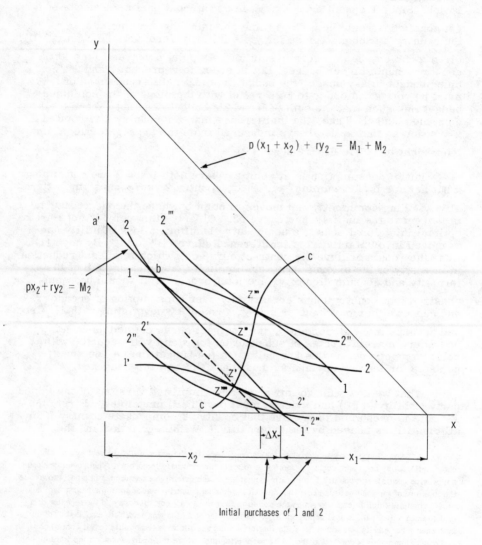

160

rule specified, the point z' on the contract curve in Figure 2 is achieved. Note that if the TP rule is specified, damages cannot occur or must be paid in kind, i.e., reduced noise only, then point a will be achieved once again but at a loss to at least one, if not both countries. Also, a TP rule which specified that the damaged country must be made better off while the liable country cannot be better off, will result in negotiation to the contract curve. The ultimate point achieved on the contract curve in this case will be to the left of point z' in Figure 2 at point z'''', i.e., x_1 will be higher while y and x_2 will be lower than for the first PP rule.

We next turn to the case of VP rule. By VP rule, we mean that there is a rule which specifies that country 2 by inadvertently creating an external diseconomy is not required to compensate country 1 for damages. Also, there are assumed to be no provisions contained in this rule which would impede private negotiation between affected countries. Clearly, in this case a negotiated settlement would involve a payment from country 1 to country 2. Costless negotiation would result in a distribution of x and y on the contract curve, both countries 2 and 1 are made no worse off and perhaps better off than at point b.

A TP rule, as specified here leads to a negotiated settlement on the contract curve cc on or to the left of point z', while a VP rule leads to a negotiated settlement on or to the right of point z''. Thus, the distribution of wealth between countries is altered by adoption of a TP or VP rule which changes the consumption patterns for x and y. Since negotiation leads to attainment of the contract curve regardless of rules, Coase is correct in asserting short-run efficiency.[1] The above derivation can be taken as a proof of the "Coase proposition" amended as follows: regardless of whether a TP or VP rule is adopted, "perturbations" arising from non-market forces can be resolved through negotiations of the countries affected, and provided transactions costs are zero after perturbation, Pareto efficiency is attained in the short run and in the long run if consumers or factors of production are immobile internationally.[2] In Appendix I, a simple model of international production is developed which demonstrates that under several rather general assumptions neither the TP or VP rules will achieve global production efficiency without other internationally established controls. The assumptions include certain types of positive transactions costs and that countries are essentially price takers. A second assertion which becomes obvious in the diagrammatic analysis is that with more or less continuous perturbations and negotiation, a VP rule will have "wealth" effects toward making some countries worse off. Not unlike gambling, there will be some countries that emerge as winners and others as losers. Since the perturbations are unforeseen in this world, it would appear Paretian in spirit to adopt a TP rule.

Some Brief Notes on Bilateral Negotiation and Game Theory

Whether the "third party must pay", "victim must pay" or some intermediate principle is implicitly or explicitly adopted, there are

1. R. Coase, "The Problem of Social Cost," Journal of Law and Economics (August 1960).
2. See Appendix I for a discussion of the possibilities when transaction costs are positive.

potential gains from trade via bilateral negotiation. Both the emitter and receptor have incentives to undertake negotiation.[1] Thus, in terms of game analysis, unadjusted externalities without threats of conflict are essentially the form of co-operative or Nash types of games. Both countries can potentially gain by co-operation although each may decide that it is in its best interest not to co-operate, or pretend not to and thereby possibly achieve a greater expected threat payoff or value. With externalities generally, it can be expected that side payments will be made in order to achieve co-operation. Thus, the classical Nash co-operative solution without side payments appears to be inappropriate. Side payments might enter for example if the upstream and downstream control costs were different. In order to illustrate some basic problems involved in negotiation for transnational externalities, we shall resort to a simple example. Let it be assumed that there are two countries 1 and 2 where 2 is the emitter and 1 the receptor. Downstream damages are 10 prior to control and 2 with control, with upstream gains due to savings in pollution control of 4. Downstream control costs are assumed to be 6. There are two decisions for each country that are not mutually exclusive, namely whether to joint the coalition and at what level to participate, i.e., how much control to provide. We shall concentrate on the decision whether to co-operate or not presuming there is no double layered process of irrevocable agreement to negotiate which is followed by negotiations. Also, it is assumed for the moment that both announce their intention to negotiate simultaneously with offers. Finally, neither country is assumed to precisely know the other's control costs and the emitter country does not accurately know the level of downstream damages.

Given the previous assumptions and a set of expectational values of each country, the maximum bid and expected bid of the receptor country and minimum acceptable offer and expected offer for the emitter country can be established. To make things simple, we shall set the following probabilities:

a) Probability that damages avoided in country 1 is 10 equals .2 in country 2 and the probability is .8 that damages avoided by control equals 2 in country 2.

b) Probability that 1's costs of control are either 4 or 6 is .5 in country 2.

c) Probability that 2's costs of control are either 4 or 6 is .5 in country 1.

Presuming the "victim must pay" principle is in effect, the maximum bid that I would make to 2 would be initial damages less damages after control which is 8. The minimum acceptable offer for 2, of course, is 4. These amounts are in effect the "threat payoffs" of the Nash type of game for coalitions. Thus, there are clearly gains from negotiation if negotiation can be initiated. However, note that "expected rent" beyond payment for control by country 2 from country 1 prior to negotiation is negative, i.e., rent defined as 2's expected value of downstream damages reduced is $[.2(10) + .8(2)]$ less 2's expected value of 1's control costs

---

1. That is, assuming the externality is continuous over time, has inadvertently arisen, and the emitter, if he does not pay, must cease the activity-generating externalities.

$[.5(6) + .5(4)]$. Thus, 2 may not undertake negotiations due to the fact that it expects 1's offer will be less than its control costs.[1]   Alternatively, 1 will decide to attempt negotiation with 2 because its expectation of gain would be $[8 - .5(4) - .5(6)]$ which exceeds its gain of 2 from instituting control unilaterally.

In Figure 3, the possible bargaining positions and "threat payoffs" for this problem are depicted. Any point on SS' is Pareto efficient in that all gains from negotiations on the transfrontier diseconomy between countries 1 and 2 are exhausted. Which point on SS' that is ultimately selected via negotiation depends on the rules established for negotiation, i. e. , once-and-for-all bid by the receptor, a sequence of simultaneous bids and demands, or sequential bids and demands. Also, the point depends on the exact specifications of the VP or TP rules. For example, if a TP principle is adopted which specifies the emitter country cannot be any better off than before the emergence of the transfrontier externality, point S will be the negotiated solution. Alternatively, if under the VP principle the receptor country cannot be made better off, then S' will be selected. Finally, if transactions-negotiation costs are introduced, then it is conceptually possible for country 2 to decide not to negotiate under either the TP or VP principles, since the country a priori perceives that it would be better off without it. (See Appendix I for an elaboration on this point.)

This simple example of a co-operative game between two countries on transfrontier externalities underscores several points. First, the bidding process itself, i. e. , who bids first and how binding is the bid, may significantly alter the outcome. In the example above, country 2 will not establish negotiations or make the first bid because expected gains are negative. Second, precise information on damages and control costs by both countries will reduce the risk associated with bidding too high to setting payments too low. Under the "TP" principles, this risk factor is intensified since the emitter country now confronts uncertainty with respect to both receptor damages and downstream control costs. Alternatively, under the "VP" principle the country making the payment, i. e. , receptor, confronts uncertainty only with regard to the emitter's control costs. Such a reduction of uncertainty on information does not occur in OECD's "polluter pays" principle where the emitter is uncertain on damages and downstream control costs and the receptor is uncertain as to upstream control costs.

CONCLUSIONS

The major conclusions of this paper are:

1.   Transnational externalities in production or consumption are involuntary transfers of perceived wealth among emitter and receptor countries. Given the general acceptance of the national sovereignty principle, such externalities of a significant magnitude will in general only

---

1.   It should be pointed out that this grossly simplifies expectation problems in that we omit consideration of country 2 calculating country 1's expectations and likewise country 1 calculating country 2's expectations on country 1, ad infinitum. Here we only consider country 1 and 2's expectations directly.

Figure 3

EFFICIENCY FOCUS FOR TRANSFRONTIER
EXTERNALITY BETWEEN TWO COUNTRIES

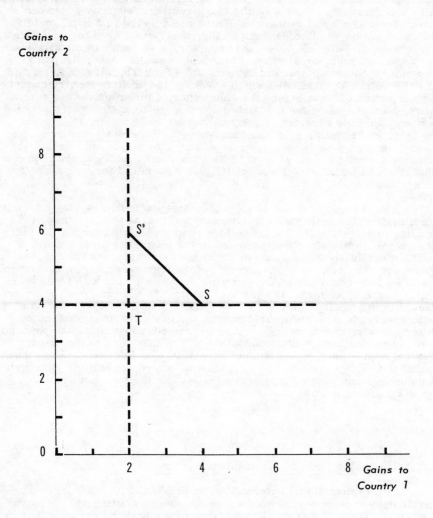

be resolvable via bilateral negotiation with side payments, usually from receptor to emitter.

2.   Unadjusted transnational externalities, if their resolution through compensation or other means shifts international prices, may make a single emitter or receptor nation better off. Thus, the link between wealth-income and welfare of a country is broken. In order for the receptor nation to be made better off, international prices must be significantly affected and willingness to pay for the commodity, influenced by the externality, must be strong if its relative price increases as a result of compensation or weak if its relative price decreases resulting from compensation.

3.   Transnational externalities, unlike domestic externalities in production, will generally not cause a shift from concavity to convexity of domestic production functions. This is due partially to the underlying assumption that resources are mobile nationally but not internationally.

4.   A "victim must pay" principle or a "third party" principle will lead to an inefficient long-run allocation of resources between countries, provided resources are mobile between them and certain assumptions on positive transaction costs are valid. [1]   These assumptions are that all firms only respond to realized profits in international location decisions and are price takers. To achieve global efficiency in resource allocation, some other types of controls need to be implemented such as restrictions on international location. The same inefficiencies might arise for transnational consumption externalities unless consumers are not internationally mobile.

5.   Transactions costs in negotiating for externalities are shown to markedly affect the expected outcomes of bilateral negotiations. If such costs are high, countries are generally risk averters, and these costs must be borne by the receptor country under the "third party" principle, then it can be expected that pollution will be greater than is desirable from a global viewpoint. There appears to be a basic asymmetry in incentives for negotiation of externalities between the "victim must pay" and "third party" principles, when transactions costs are positive. Under the "VP" rule, both emitter and receptor have incentives to negotiate and thereby incur transactions costs. Alternatively, with the "TP" rule, only the receptor has such incentives since the emitter is always better off by not negotiating.

6.   Unidirectional transnational externalities were shown to be conceptualizable as a special case of co-operative games under the "victim must pay" principle. It was argued that rules on negotiation may substantially alter gains between countries undertaking bilateral negotiation for transfrontier externalities. Thus, rules on negotiation and arbitration that are equitable should be developed and advocated by international agencies.

1.   See Appendix I for a semi-rigorous proof.

Appendix I

A NOTE ON TRANSFRONTIER EXTERNALITIES, TRANSACTIONS COSTS, AND LONG-RUN INTERNATIONAL ADJUSTMENT IN FIRMS AND FACTORS OF PRODUCTION [1]

by

R. d'Arge and W. Schulze
University of California, Riverside

There appear to be four cases regarding transactions costs for transfrontier externalities:

a) where transactions costs are always or nearly zero between emitter and receptor countries;

b) where transactions costs are positive and significant both before and after emergence of an externality;

c) where transactions costs are positive before the externality appears but zero thereafter; and

d) where transactions costs are zero before the externality appears but positive and perceptively significant thereafter.

Case a) can easily be disposed of as one which rules out the existence of externalities that are not a priori resolved by market negotiations. The Coase proposition is a special case of c). Case d) appears to be unreasonable. Finally, case b) is the important one taxonomically for analyzing "real world" problems. An important subset of cases under case b) arises where transactions costs are different for the two parties either independent of or dependent upon the prevailing rule for liability, i. e. either the "third party," "polluter pays", or "victim pays" principle are operative or not.

Transactions costs may affect negotiations in a multitude of ways depending on their source. These include: uncertainty and information gaps (or costs); known or unknown contracting or negotiation costs; costs associated with organising and sustaining negotiations between countries including dissemination of information; and enforcement costs for existing contracts or treaties. Of these different types of transactions

---

1. A model not too dissimilar to the one applied here emphasizing intra-country location is developed by W. J. Baumol in "On Taxation and the Control of Externalities, " American Economic Review (June 1972). This is an abbreviated and transformed section of a paper by the same authors entitled, "Coase Proposition, Wealth Effects, and Long Run Equilibrium, " Working Paper No. 19, Program in Environmental Economics, University of California, Riverside (April 1972).

costs we shall concentrate briefly only on two types - those associated with confronting an uncertain prospect of a future perturbation (externality) and those costs associated with negotiation once the externality has occured. What we wish to establish is that differences in risk aversion between emitters and receptors will cause adjustments in the allocation of resources and negotiated level of externality depending on whether the "third party" principle, or "victim pays" principle prevail.

A realistic case is one where, under complete liability, the receptor country incurs negotiation costs and, under complete non-liability, the emitter country must pay such costs. Thus, those who potentially gain are assumed to initiate negotiation and underwrite the cost of negotiation. Let ψ_L^*, ψ_{NL}^*, ψ_L, and ψ_{NL} denote the level of externality-generating activity, for the cases of zero transactions costs and TP, zero transactions costs and VP, positive transactions costs with TP, and positive transactions costs with VP rules, respectively. Given the assumption that the contract curve is upward sloping, and not too dissimilar marginal utilities of income or marginal disutilities of payments for negotiation exist between countries, then it can be asserted $\psi_L^* \lesseqgtr \psi_L$ and $\psi_{NL}^* \lesseqgtr \psi_{NL}$. That is, positive negotiation costs, regardless of VP for TP rules, will impede negotiation so that the optimal level of externality-generating activity with zero negotiation costs is never achieved. Given the assumptions above, then $\psi_L^* \lesseqgtr \psi_{NL}^*$.

This implies $\psi_L^* \lesseqgtr \psi_{NL}^* \lesseqgtr \psi_{NL}$ but does not imply $\psi_L \lesseqgtr \psi_{NL}$. Thus, costly negotiation where one country incurs these costs may lead to a case where the TP rule results in a higher level of externality-generating activity than the VP rule. This outcome can be induced by differences in marginal utility of income between the emitter and receptor countries as well as a large number of other assumptions on initial endowments or preference maps. The important point here is that it cannot be a priori determined that complete liability or the "third party" principle will reduce external diseconomies by a greater amount than no such principle when negotiation costs are introduced and must be paid by the country initiating negotiation. If negotiation costs are different for the two countries, the outcome is even less clear. It has often been contended that emitters must have lower negotiation or organisation costs than receptors since receptor countries are generally more in number while the emitter is viewed as a single country (source). With this type of differential in negotiation costs, there is still no clear-cut statement that can be made on the inequality between ψ_{NL} and ψ_L. It depends again on who incurs the organizational or negotiation costs. Under the TP rule, the receptor country must undertake negotiation costs since there is no incentive for the emitter to do so. However, under the VP rule, there is an incentive for both to undertake negotiation and incur such costs. If the receptor must pay negotiation costs under either legal rule, then without further assumptions the inequality between ψ_{NL} and ψ_L cannot be determined. If the receptor country pays negotiation costs with the TP rule but the emitter country pays these costs under the VP rule, then it can be expected that $\psi_{NL} \lesseqgtr \psi_L$, provided the emitter has lower negotiation costs. What is important from the above statement is that the "third party" principle (TP) or lack of it with negotiation costs for allocative efficiency requires an additional rule specifying who incurs these costs. Without such a rule, negotiation may be completely stopped and thereby yield inefficiencies.

Thus far, we have not introduced uncertainty explicitly into the discussion even though the characterization of a world with externalities hints at analyzing externalities as unexpected, or at least, uncertain events. If externalities can be identified as uncertain events where a probability distribution is identified for each type of externality and there are methods to reduce the probability of occurrence to zero or some reasonable level, e. g. , construct and operate a tertiary treatment plant on the Rhine, then the externality problem can be analyzed, at least partially, with tools from probabilistic microeconomics. We shall not do that here but suggest some obvious results. First, if the emitter country is more risk averse than the receptor, then its government may require purchase control devices under the TP but the receptor may not with its absence.

So far, we have viewed transactions costs as the dominant factor in externality negotiations. Next, we turn to a semi-classical long-run case of competition between firms, in an international context. In so doing, a particular set of transactions costs are preserved, namely, that firms are price takers, but output adjustments shift international prices. Firms are assumed to move internationally in search of the highest profits with no hindrance by governments or entry costs and observing no other signals than current profits. Thus, a "perfect capital" market internationally is implied. We also make the simplifying assumption that each country has a distinct comparative advantage in producing one type of commodity and that is all it produces. In addition, to avoid balance of payments and other complications, we presume that each country is in a partial equilibrium world where demand price for their product is prespecified at any level of output. Finally, no third party is presumed to enter and arbitrage externalities such that all potential gains from trade are exhausted. The question to be resolved is whether or not negotiations between producing countries' firms exploiting the gains from trade made available by an externality in production will result in an efficient solution no just in the short run but in the long run among countries. Since profitability among countries determines entry (presumed to be costless), and since the adoption of the TP principle with respect to externalities will affect relative profitability in emitter and receptor countries, it is clear that the international location of industry will be affected by which principle is adopted. Free entry may be inconsistent with stability of a negotiated solution arising under a non-TP principle because new entrants can act as free riders, a possibility excluded by the TP rule or a Pigovian tax. It is assumed that firms perceive prices as constants. We also assume that only those directly involved will negotiate.

We will consider a two-country partial equilibrium model with a diffuse externality such that the output of industry in country 2 adversely affects the production of every firm in country 1 in a like manner. Each country is assumed to produce one product under competitive international pricing. An example might be the release of air pollutants by industry of one country into an airshed with instantaneous horizontal mixing which also contains a second country's industry. The question would then be to determine how large the industries should be in each country. Where there are n_1 identical firms each producing output y_1 in country 1 and n_2 identical firms each producing output y_2 in country 2, we can write inverse demand functions (price P_i as a function of country output $n_i y_i$) for country 1 as $P_1(n_1 y_1)$ where $P_1'(n_1 y_1) < 0$ and for country 2 as $P_2(n_2 y_2)$ where $P_2'(n_2 y_2) < 0$, where prime denotes a first derivative. Note that each country is under competition, so it takes prices as given. The total production cost for each firm in country 1 is given as $C_1(y_1) + D_1(n_2 y_2)$ where $C_1(y_1)$ is the direct cost of production for each firm in country 1 and $D_1(n_2 y_2)$ is the damage incurred by each firm in country 1 as the result of the emissions of country 2. Note that we are presuming damages are separable. However, our results concerning the optimality of various policy measures are generally not dependent on this assumption. Total cost for each firm in country 2 is $C_2(y_2)$. We assume C_1', C_2', C_1'', $C_2'' > 0$ and D_1', $D_1'' > 0$ in the relevant regions of production.

The conditions for a global optimum for both countries taken together are generated by maximizing net benefits (NB) which can be defined as the difference between willingness to pay for the output of both countries and the total cost of production in both countries. Thus, the international optimum is a maximum of:

1) $$NB = \int_0^{n_1 y_1} P_1(s_1) ds_1 + \int_0^{n_2 y_2} P_2(s_2) ds_2$$

$$- \left[n_1 (C_1(y_1) + D_1(n_2 y_2)) + n_2 C_2(y_2) \right] \qquad y_1, y_2, n_1, n_2 \geqq 0$$

where s_1 and s_2 are dummy variables of integration in the demand functions for the output of each country. Assuming an interior solution, the first order conditions are:

1.1) $\partial NB / \partial y_1 = n_1(P_1 - C_1') = 0$

1.2) $\partial NB / \partial y_2 = n_2(P_2 - C_2' - n_1 D_1') = 0$

1.3) $\partial NB / \partial n_1 = P_1 y_1 - (C_1 + D_1) = 0 = \pi_1^*$, and

1.4) $\partial NB / \partial n_2 = P_2 y_2 - (C_2 + n_1 D_1' y_2) = 0 = \pi_2^*$.

The interpretation of (1.1) and (1.2) is quite straightforward and implies, where n_1, $n_2 \neq 0$, that for each firm in country 1 price should be equal to marginal cost (C_1'), and that for each firm in country 2 price should be equal to marginal cost (C_2') plus marginal damages to country 1 ($n_1 D_1'$). These are the usual short-run conditions with a unidirectional externality between countries. The conditions for a long-run optimum, (1.3) and (1.4), are more interesting since they should correspond to the definition of zero profits for firms in countries 1 and 2 respectively (assuming firms enter or leave countries until profits are zero). Equation (1.3) implies that $\pi_1^* = 0$ is the optimum level of profits where the receptor country bears the full cost of the externality D_1 at the optimum. This result suggests that compensation for damages will distort long-run equilibrium in the receptor country. Equation (1.4) implies that $\pi_2^* = 0$ is the optimum level of profits where the emitter country must bear an additional cost of $n_1 D_1'$ per unit of output y_2 produced. This can be interpreted as an optimum long-run Pigovian tax equal to marginal damages on the output of the firms in the emitter country.[1] We note then, that the optimal policy barring renegotiation after taxation by an "international tribunal" or agency is to do nothing with respect to the receptor country, allowing it to bear the cost of the externality after the optimal tax on output has been applied to firms in country 2. This will assure the optimum number of firms in each country.

The relationship between the optimal Pigovian tax case (denoted *), the unadjusted externality case (denoted E), and the TP principle case (denoted P) can be best demonstrated with the aid of Figures 1 (a firm in country 1) and 2 (a firm in country 2). In Figure 1, the optimal long-run equilibrium point for receptor firms is * at the lowest point of the average total cost curve including optimal damages suffered (AC_1^*).

This point is defined by the intersection of the marginal cost curve (C_1') with the adjusted average total cost curve (AC_1^*). As damages (D_1) increase with increasing output of country 2, the average total cost curve of firms in country 1 including damages shifts upward. We presume that where the optimum tax is applied to firms in country 2 and free entry exists for both countries, there will be a convergence to D_1^* the optimal long-run level of damages, and optimal price P_1^* and quantity y_1^* will result from the long-run equilibrium point (*). Note that this optimum is a basis of comparison since gains from trade are possible between the two countries so long as a Pigovian tax is not charged and randomly distributed.

The optimum equilibrium point in Figure 2 for firms in country 2 is also denoted *. This can be defined by the intersection of the average total cost curve including the tax (AC_2^*) with marginal social cost (C_2' + $n_1 D_1'$). Note that this point corresponds to the zero profit condition for firms in country 2 where AC_2^E is the unadjusted average cost and the area P_2^E E * P_2^* is the optimal long-run tax collected from each firm in the emitter country. This implies that if through some mechanism not involving a tax or levy the two countries reach the optimum points * in Figures 1 and 2, positive profits equal to the area P_2^E E * P_2^* times n_2^* will be obtained. Since positive profits will induce more firms to enter country 2, the optimum point * cannot be a stable equilibrium under free entry. Thus, the Pigovian tax, if achieved through international management agencies, serves to remove these destabilizing profits.

---

1. A similar result is obtained by Baumol in his American Economic Review article. Baumol, op. cit. (June 1972).

Figure 1

COSTS AND OUTPUT ADJUSTMENTS
OF ONE FIRM IN RECEPTOR COUNTRY

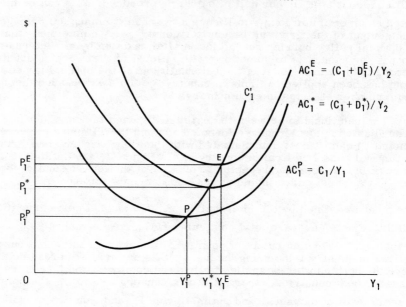

Figure 2

COSTS AND OUTPUT ADJUSTMENTS
OF ONE FIRM IN EMITTER COUNTRY

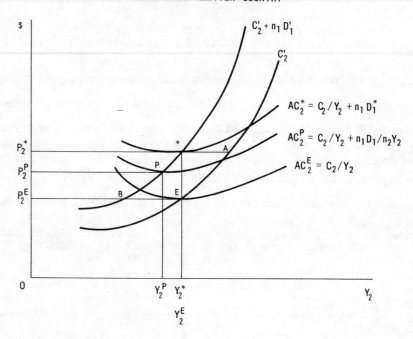

The uncompensated externality case results in a long-run equilibrium at point E in Figure 2 for firms in country 1. Here, since damages received (D_1^E) will be greater than optimal (D_1^*), the average cost curve (AC_1^E) will lie above AC_1^*. Since the externality is separable, the marginal cost function (C_1') does not shift, so entry or exit occurs until profits are zero resulting in a long-run price of P_1^E and output per firm of y_1^E. This implies, where the demand function for country 1 as a whole is downward sloping ($P_1' < 0$), that total industry output and number of firms will be less than optimal in the unadjusted externality case for receptors, since externality price and output for each firm are both greater than optimal price and output per firm, respectively. This occurs because at the higher (non-optimal) price, demand for the country's product is less. In Figure 2 the long-run externality equilibrium point for firms in the emitter country occurs where average cost (AC_2^E) equals marginal cost (C_2') resulting in a price (P_2^E) lower than the optimum price (P_2^*). However, firm size is still optimal since output (y_2^E) in this case is identical to the case under taxation (y_2^*). The intuitive explanation of this result which is not dependent on separability is simply that, in spite of the externality, international product is still maximized by producing each unit of y_2 as cheaply as possible. This implies that in the unadjusted externality case, there will be too much total output from the emitter country and too many firms, because the demand curve for the country is assumed to be downward sloping ($P_2' < 0$) even though each firm perceives demand as infinitely elastic. Thus, in the unadjusted externality case, there is an under-allocation of resources to the receptor country and an over-allocation of resources to the emitter country.

In the complete liability or TP case discussed here we assume that firms in country 1 are compensated for damages and that potential entrants into the industry of country 1 are aware that they too will be compensated. Firms in country 2 are responsible for damages done to country 1 and we assume, since the externality is diffuse and the firms are taken as identical, that each must bear the cost of compensation equally. With liability, profits for firms in country 1 and 2 can then be written as:

2) $\quad \pi_1 = P_1 y_1 - C_1(y_1) - D_1(n_2 y_2) + [D_1(n_2 y_2)]$ and

$\quad \pi_2 = P_2 y_2 - C_2(y_2) - [n_1 D_1(n_2 y_2)/n_2]$

where the terms in brackets are compensation or liability payments by firms in each country respectively. The first order conditions for maximum profits in each firm, assuming an interior solution, are:

2.1) $\quad \partial \pi_1 / \partial y_1 = P_1 - C_1' = 0$ and

2.2) $\quad \partial \pi_2 / \partial y_2 = P_2 - C_2' - n_1 D_1' = 0,$

which imply that the conditions for short-run optimality are satisfied. However, if it is assumed that firms enter until profits are zero, compensation to country 1 results in a long-run equilibrium (position P in Figure 1), the lowest point on the average cost curve without damages (AC_1^P) for firms in the receptor country. Again, the marginal cost function $(C_1^!)$ does not shift since damages are separable and the resulting price (P_1^P) and output for each firm (y_1^P) under the TP principle are less than optimum. Thus, both total country output and the number of firms will be too large for the receptor country with the TP principle case since demand for the country's output will be greater at the lower price.

Turning to Figure 2, firms in country 2 will reach an equilibrium point (P) in the long run under the TP principle which is the lowest point of the average total cost curve including each firm's share of damages to be paid (AC_2^P). Note that because total damages increase at an increasing rate, marginal damages $(n_1 D_1^!)$ are greater than average damages $(n_1 D_1/n_2 y_2)$, so AC_2^P lies below AC_2^* and therefore the intersection of the marginal social cost function $(C_2^! + n_1 D_1)$ and AC_2^P must be below and to the left of the optimum point (*). This is the point of zero profits including liability for damages for the emitter country. Both price (P_2^P) and output (y_2^P) are too low for each firm in country 2. However, total country output will be too high and there will be too many firms in the emitter country under a TP principle since, given the lower price, aggregate demand for the country's output will be too high. Thus, there results a long-run over-allocation of resources to both country 1 and 2. Taxes or other controls are necessary to prevent a misallocation of resources in long-run international production even if there are well-defined rights with a TP principle on the externality. A TP principle solution could be adjusted to the optimum equilibrium point by taxing receptors an amount equal to the damage payments they receive and taxing emitters an amount equal to the difference between average damages and marginal damages, a procedure inefficient as regards international information and enforcement costs to make this bargaining solution appear unattractive.

The complete non-liability case or victim-must-pay principle can best be explained in two steps. First, we will demonstrate that a negotiated solution under a non-liability rule cannot sustain the optimum points (*) assuming that the number of firms in each country is constrained to be less than or equal to the optimum $(n_1 \leqq n_1^*, n_2 \leqq n_2^*)$. As will be seen later, this assumption prevents a free rider problem from upsetting the potential equilibrium point (*) in Figures 1 and 2. Assume that firms in both countries are initially at *. Next, observe that * for firm 1 in Figure 1 is a point of zero profits. However, in Figure 2 it is clear that firms in country 2 could earn profits greater than those obtained under * by moving to point A. Thus, for * to be achieved by negotiation for firms in country 2, firms in country 1 must offer to pay a bribe at least equal to the difference between profits at * and profits at A to existing firms in country 2. Clearly, firms in country 1 are making zero profits at * in Figure 1 and cannot pay any bribe. Thus, the optimum points * in Figures 1 and 2 are not feasible under a victim pays rule even ignoring the destabilizing effects of entry on coalitions since firms in country 2 must be made at least as well off at market price P_2^* as they as they would be by not adjusting for the externality. Thus, firms in country 2 would be unwilling to remain at *. It is conceivable, with the number of firms fixed by controlling entry through some licensing or nationals' ownership process, to achieve a short-run optimum with a solution somewhare along the marginal social cost function $(C_2' + n_1 D_1')$ in Figure 2. However, the number of firms in country 1 must be fixed $(n_1 \leq n_1^*)$ such that profits sufficient to cover bribes to country 2 can be obtained. Clearly, without some taxation policy, even by controlling entry, the long-run optimum solution is not attainable under the VP principle since the receptor country cannot afford to bribe emitters.

If free entry is allowed, potential entrants always have a valid threat of entry in the non-liability case if market prices are above P_2^E in country 2 or above the lowest point of the current average cost plus average damages in country 1. It is clearly impossible to bribe potential firms to stay out of a country as long as they could earn positive profits by entering, because under perfect information regarding current profitability, an indefinitely large number of potential firms would eventually threaten to enter. In Figure 2, entry would result in an eventual price of P_2^E for firms in country 2 with a short-run optimum position at point B, implying negative profits for emitter firms. However, this point cannot be stable because firms in country 1 must earn sufficient profits to compensate firms in country 2 for their losses. But free entry into country 1 will tend to force profits in that country to zero by a free rider process where receptors will enter (given a level of emissions reduced by negotiations between existing firms), join the coalition, but find profits eventually reduced to the point where firms in country 2 can no longer be bribed to reduce output. Note that this process assumes that firms are price takers, i. e., perceived demand for each firm or potential entrant is infinitely elastic. Thus, potential entrants cannot realize that entry must lower prices, thereby destabilizing existing solutions; nor does the country's government establish entry constraints by assumption. This sketch of events implies that under a VP rule with free entry and negotiated solutions is unstable. One can again imagine a sufficiently complicated set of regulations and/ or taxes to allow an optimal solution to be obtained under a VP rule.

Although we have excluded explicit negotiation costs from this analysis, it is difficult to imagine any set of circumstances, in which firms are competitive (free entry is of course a necessary condition for the existence of competitive firms), an externality exists between countries, and long-run optimality is achieved without taxation or other forms of internationally established entry restrictions. It is possible that governments could act in concert to simulate a "Pigouvian" solution provided agreement could be achieved on the "third party" principle. In this case, an efficient allocation of global resources could be achieved by the receptor government collecting damage payments from the emitter government and using them for purposes other than compensating their firms or citizens adversely affected.

A BASIN AGENCY AND THE JOINT EXERCISE
OF SOVEREIGN POWERS

by

Herbert A. Howlett

Delaware River Basin Commission
Trenton, N. J. , U. S. A.

Few examples exist of agencies which have been created for implementing policies on international transfrontier pollution. However, an agency has been in operation within the United States for some ten years which has bestowed upon it the joint exercise of some of the sovereign powers of four states and the federal government, among them the authority to abate pollution at any location within the area of its jurisdiction. That agency is the Delaware River Basin Commission.

A review of the powers and performance of the Delaware River Basin Commission may be useful in considering an international agency to implement policies to manage transfrontier pollution.

The Delaware River Basin

The Delaware River Basin occupies a position on the Atlantic Coast of the United States about midway between New York City and Washington, D. C. The Delaware River and its tributaries drain a watershed of approximately 12,750 square miles. In its 320 mile length, the river passes through or serves as the border of the States of Delaware, New Jersey, New York, and Pennsylvania. The present population of the basin is in excess of 7,000,000 , its largest city being Philadelphia, Pennsylvania.

The uppermost reaches of the Delaware River watershed are used for agricultural and recerational pursuits, and the streams are of a quality which support and propagate trout even though carrying a worrisome nutrient load. The City of New York exports nearly one-half of its water supply from this portion of the watershed. The lower portions of the river and Delaware Bay (a reach of about 140 miles) are tidal and heavily used for navigational purposes, accommodating drafts up to 40 feet. The port area, extending from Trenton, New Jersey to Wilmington, Delaware, is significantly polluted by the population and industrialized area of Philadelphia. In this polluted area, the Delaware River serves as the common boundary between New Jersey and Pennsylvania, and then flows into Delaware before discharging into Delaware Bay, which is a common boundary of New Jersey and Delaware.

The Delaware River Basin Compact

After years of indecision, legal battles over exportation of water from the upper river, and worsening pollution in the lower river, a disastrous flood in 1955 stimulated the governors of the four basin States and the mayors of the Cities of New York and Philadelphia to constitute a Delaware River Basin Advisory Committee (DRBAC) charged with the duty of recommending a form of interstate agency empowered to manage the water resources of the basin. About this same time, the Delaware

River Basin Research, Inc., entered into a contract with the Syracuse University Research Institute "for a study of 'governmental organisations for development of the Water Resources of the Delaware River",[1] with Roscoe C. Martin as study director.

Thus, the principles contained in the Delaware River Basin Compact[2] were born. In less than one year, these principles were approved by four state legislative bodies and governors, and by the Congress and President of the United States. In so doing, each of the five signatory agencies made possible the joint exercise of sovereign powers over the water resources of the basin in the common interests of the people of the region.

In the Compact's opening sentence, "the signatory parties recognize the water and related resources of the Delaware River Basin as regional assets vested with local, state and national interests, for which they have a joint responsibility ... ", and proceeds to establish the Delaware River Basin Commission and the authority and principles under which it is to govern. In the words of the Compact /Sec. 1.3(e)/ - "In general, the purposes of this compact are to promote comity; to remove causes of present and future controversy; to make secure and protect present developments within the states; to encourage and provide for the planning, conservation, utilization, development, management and control of the water resources of the basin; to provide for co-operative planning and action by the signatory parties with respect to such water resources; and to apply the principles of equal and uniform treatment to all water users who are similarly situated and to all users of related facilities, without regard to established political boundaries. "

The Compact was drawn for an initial period of 100 years, with only the federal government having the right of withdrawal, and that only by Congressional action.

The area to which the authority given in the Compact applies is the area of drainage into the Delaware River and its tributaries, including Delaware Bay.

Anticipating problems of co-ordination and responsiveness, the Compact imposes specific requirements upon the signatory agencies and their subdivisions in Article 11, upon the general public of the basin in Section 3.8, and upon its Commission in Section 1.5.

Article 11 provides that (a) the planning of all water related projects sponsored by federal, state and local agencies shall be undertaken in consultation with the Commission, and (b) no expenditure or commitment shall be made by any of these governmental agencies for or on account of construction, acquisition or operation of a project unless it has been included by the Commission in its Comprehensive Plan.

Section 3.8 provides that:

"No project having a substantial effect on the water resources of the basin shall hereafter be undertaken by any person, corporation

1. Martin, Roscoe C. et al, River Basin Administration and the Delaware, Syracuse University Press (1960).
2. Delaware River Basin Compact, United States Public Law 87-328, Approved September 27, 1961, 75 Statutes at Large 688, Trenton, New Jersey, 1961.

or governmental authority unless it shall have been first submitted to and approved by the commission. "

Finally, Section 1. 5 provides that:

"It is the purpose of the signatory parties to preserve and utilize the functions, powers and duties of existing offices and agencies of government to the extent not inconsistent with the compact, and the commission is authorized and directed to utilize and employ such offices and agencies for the purpose of this compact to the fullest extent it finds feasible and advantageous. "

The Delaware River Basin Commission

The Compact creates the Delaware River Basin Commission, hereinafter referred to as the Commission,

"as a body politic and corporate, ... as an agency and instrumentality of the governments of the respective signatory parties. "

The Commission consists of the governors of the signatory States and one commissioner appointed by the President of the United States. Each of these commissioners may appoint an alternate to act in his stead, but the alternate may not redelegate his voting privilege. Each commissioner is entitled to one vote, and except in matters relating to the annual budgets, a simple majority vote is ruling. Annually, the Commission elects its chairman and vice-chairman, and historically, these offices have rotated among the signatory agencies.

Powers and Duties of the Commission

Among the duties of the Commission, mandated by the Compact, are the adoption of (a) a comprehensive plan for the immediate and long range development and uses of the water resources of the basin, (b) a water resources programme based upon the comprehensive plan highlighting for the short-term (six year) i) the relationship between water supply and demand, ii) the projects required to meet the demand and iii) the projects proposed to be undertaken by the Commission, and (c) annual current expense and capital budgets. Prior to adoption of any of these documents, the Commission is required to duly notify the public of the pending action and provide for public hearings on the issues.

The Commission is given the power to allocate the waters of the basin in accordance with the doctrine of equitable apportionment, and to impose conditions with regard to such allocations. The Commission may (a) sue or be sued in a court of competent jurisdiction, (b) borrow money, (c) issue revenue bonds, and (d) acquire land for authorized projects by condemnation proceedings. It may not (a) interfere with the laws of signatory parties relating to riparian rights, (b) undertake construction of physical facilities until authorized to do so by separate action of the Congress of the United States unless no federal funds are to be utilized, (c) charge for water that could have been taken lawfully without charge prior to the Compact date, or (d) impose any charge for commercial navigation (jurisdiction reserved to the federal government).

Specific authority is given to the Commission with regard to pollution control. Under the General Powers (Sec. 5.1) the Compact provides:

"The Commission may undertake investigations and surveys, and acquire, construct, operate and maintain projects and facilities to control potential pollution and abate or dilute existing pollution of the water resources of the basin. It may invoke as complainant the power and jurisdiction of water pollution abatement agencies of the signatory parties."

The Commission may adopt standards governing the quality of water within the area of its jurisdiction. It may also, after due legal process, issue an order upon any person or public or private corporation, or other entity, to cease the discharge of waste waters which are in violation of its rules and regulations. Anyone who violates any provision of the Compact or any rule or regulation duly promulgated thereunder,

"shall be punishable as may be provided by statute of any of the signatory parties within which the offense is committed, provided that in the absence of such provision any such person ... shall be liable to a penalty of not less than $50 nor more than $1,000 for each such offense to be fixed by the court which the commission may recover in its own name in a court of competent jurisdiction ... each day of violation ... shall constitute a separate offense." (Compact, Sec. 14.7)

While the Commission is empowered to develop and effectuate plans in a number of technical fields, always related to the water resources of the basin, the foregoing highlights those matters which are most relevant to considering mechanisms for implementing policies on transfrontier pollution.

An Appraisal - Delaware River Basin Compact and Commission

Point No. 1: A forum is essential to the friendly solution of any problem involving two or more parties.

The Delaware River Basin community, having encountered floods, droughts, exportations of water and legal actions thereon, and gross pollution, recognized the need for the existence of a forum composed of decision-capable members in which all matters relating to the management of the waters of the entire watershed could be discussed and resolved expeditiously.

An agency known as the Interstate Commission for the Delaware (INCODEL) endowed only with powers of persuasion, created in 1937 by the joint action of the four basin States had proved by 1955 unable to cope with problems demanding decisive action.

The need for a better forum than INCODEL led to the creation of another forum, a special task force known as the Delaware River Basin Advisory Committee (DRBAC). The basic charge of this ad hoc group was to recommend a form of regional agency capable of succeeding where INCODEL had failed. In carrying out its task, DRBAC drew upon the services of a respected institution, Syracuse University, and consultant.[1] The prestige of both the institute and the consultant were undoubtedly factors in gaining acceptance of the Compact, which was drawn from the consultant's and the DRBAC's recommendations.

Point No. 2: Pattern the charter of the basin agency to the needs of the region. Do not be needlessly restrictive, allow flexibility of administration and stimulate appropriate history to enable future generations to understand "legislative intent".

The Delaware River Basin Compact, as adopted by the five parties in 1961, has not been amended to date. This may not reflect the worthiness of the original document as much as the requirement of concurrent parallel legislation by the four States followed by consistent action by the United States Congress. Nevertheless, after ten years only minor deficiencies in the basic enabling legislation (the Compact) have been identified, none of which has had serious consequence. Most relate to the tediousness of operational procedures rather than to matters of substance.

The difficulty of obtaining an amendment to the Compact was proven when attempts were made to remove or raise the legal limit of interest on Commission bonds (6%). One of the signatory States was reluctant to grant to the Commission interest rates it had denied to its other internal agencies, even though the Commission is only authorized to issue revenue bonds, fully supported by revenue from the services rendered.

It has been necessary to seek clarification of the "legislative intent" in requiring specific Congressional approval of a project where federal funds would be used (in part) as authorized in a grant programme.[2]

Point No. 3: Select the governing body from decision-making levels (not technical levels) who have direct access to the highest authority in his jurisdiction.

The Compact names the Governors of the signatory States and a Presidential designee[3] as the Commissioners. Each Commissioner names an alternate Commissioner to act in his stead. The majority of the alternate Commissioners from the states have been cabinet-level members of the Governors', most top administrative officers of water or conservation departments. All other state appointments, whether from government or private enterprise, have been knowledgeable of State policy and have had direct access to their Governors when the situation demanded. The alternate Federal Commissioner has dedicated full time to affairs of the Commission, and he, too, has had direct access to the Presidential designee.

The position of the Commissioners and Alternates has been a source of major strength for the Commission, as it is composed of decision-makers, fully empowered to act in the best interest of his clientel. This

1. Martin, Roscoe C. et al, River Basin Administration and the Delaware, Syracuse University Press, 1960.
2. The United States Government makes non-repayable grants of money for the construction of certain pollution control facilities, in accordance with standing enabling legislation and annual appropriations to a disbursing agency.
3. Throughout the Commission's ten-year history the President of the United States has named the Secretary of the Interior, a cabinet level officer, as his designee.

has led to both timely decisions by the Commission, and joint Commission and State decisions where the matter at hand related to both governments.

Commission Administration

The Commission maintains a relatively small, but select, staff to carry out its broad mandate. This staff conducts the necessary planning and review functions, and prepares recommendations for Commission decisions at its monthly meetings. Decisions on all matters of external concern are made by majority vote of the Commission, sitting in open and publicized meetings.

The staff has entered into so-called "Administrative Agreements" with the water agencies of the signatory parties to help eliminate duplication in the review process and to provide for channels of communication. This procedure has been only partially successful, as the purposes of project reviews of any two agencies are usually slightly different, still requiring the full bureaucratic process.

Tests of the Basin Commission Mechanism

The Commission has developed, adopted and kept current a Comprehensive Plan for the management of the water resources of the basin. Responsive to the Compact (but somewhat unlike other plans), this Comprehensive Plan includes statements of policy to guide others interested in developing and using the waters of the basin, as well as a schematic plan of physical facilities which is envisioned to be required to supplement base stream flows and control floods for the benefit of present and future generations. As such, the specified uses to be made of the waters, the stream quality objectives required to make these uses possible, and the effluent waste standards are a part of this Comprehensive Plan. (The vote to approve these water quality standards and place them within the legally binding Comprehensive Plan of the Commission was four in favour and one against. Here we have an example of a vote going against one of the signatories on a major item; yet, he and that portion of his jurisdictional area within the Delaware River Basin are bound by the majority rule.)

Within the policies set forth in its Comprehensive Plan, the Commission successfully accomplished an allocation of the carbonaceous oxygen demand assimilative capacity of an 85-mile tidal reach of the Delaware River involving dischargers in three States. A generalized description of the procedures followed in this activity is found in Attachement A. It should be noted that a special committee was created by the Commission to provide advice and co-ordination in developing the detailed procedures. The legal doctrine of "equitable apportionment" was basic to the allocation technique utilized. Under the allocation procedure, the Commission imposed absolute limits, in terms of pounds of oxygen demanding waste, upon each of the dischargers. Normal growth must be accommodated within the allocated poundages; applications for supplemental allocations can be made for abnormal growth, such as a major plant expansion or an annexation of suburban districts by a city.

184

Within the present topic of discussion, the fact that an allocation was made involving three States is evidence that the joint exercise of sovereign powers has been successful in gaining agreement on the level of quality to be maintained on interstate and intrastate streams.

The Delaware River Basin was subjected to drought conditions of unusual proportions during 1964 and 1965. This drought threatened to exhaust the water supplies of the Cities of New York (dependent upon exportation of water from the Delaware) and Philadelphia (a direct user from the tidal river). The Commission, utilizing the emergency powers given in the Compact, assumed administrative direction over all remaining water stored in the basin, apportioned the supply among all, and managed to avoid a major catastrophe. This action is particularly noteworthy as it required the Commission to set aside temporarily a decision of the U. S. Supreme Court.

Again, this is an example of the merits of a forum of decision-makers who are acquainted with all aspects of the river basin, enabling them to make expeditious decisions on matters of major importance.

The foregoing tests have all indicated strength in the joint exercise of sovereign power approach. There are a number of powers provided to the Commission by the Compact that have not yet been utilized. These include:

a) purchase or construction of a physical facility that would require Commission ownership and operation;

b) issuance of bonds;

c) condemnation of property;

d) legal action, either to prevent construction or operation of a project that is in conflict with the Commission's Comprehensive Plan, or to force an existing project to comply with that Plan.

Limited use has been made of the Commission to serve as an agent of the signatory parties, but the Commission has used the services of the signatory parties in the surveillance of pollutants and for some aspects of project review. Of these aforementioned matters, two warrant special comment:

1. Ownership and Operation of Projects

Although two States operating through the Commission have agreed to share the cost of the water-supply aspects of a multiple-purpose project constructed, owned and operated by the federal government, by formulae developed by staff, (the project located entirely within the territory of one of the States), the Commission has so far not found a situation which would mandate its becoming a project owner/operator. On one occasion, a local level of government petitioned the Commission to construct and operate a major pumping project designed to divert water from one river to another watershed, but the Commission demurred. In this instance, the majority of the Commissioners expressed the philosophy that all other options for ownership, construction and operation should be exhausted before the Commission should assume these responsibilities. Even with local agency ownership, they argued, in periods of emergency the Commission could intercede and impose operational constraints. There being an option of local ownership/operation, this position prevailed after considerable debate and time had elapsed.

185

The Commission did express its willingness to become the owner, but not operator, of a regional sewerage system after alternative ownerships had been fully explored and none found to be viable. In this instance, the area to be served contained more than one jurisdiction in one State, and the outfall from the proposed waste treatment plant would have been in another State. In the final analysis, the project proved to be financially unfeasible. The time taken by the Commission before agreeing to become the project sponsor was a factor in the question of financial feasibility.

An opportunity for the Commission to become the owner/operator of a major facility is now pending. This proposed project would involve the interception and treatment of waste water, thus avoiding potential pollution of a proposed large reservoir. The area in question lies in three States (all signatory to the Compact) and numerous lower level jurisdictions. The Commission has announced its intention to be the owner/constructor of the central features of this pollution control plan, and current efforts are proceeding under that declared policy. Future decisions on this matter may be of interest to OECD as agricultural nutrients carried by the waters entering the proposed reservoir may require special remedial measures that might not be required in the absence of the reservoir. Further, the agricultural areas lie principally in an upstream State, which will not be materially benefitted by the reservoir; the reservoir would lie almost entirely in two downstream States, both of which would be substantially benefitted by the proposed project. At this point in time, one can only speculate on the outcome of this matter. It should be noted, however, that the Commission has adopted water quality standards for the upstream area, within which the local constituency must comply. There is some evidence that the Commissioners are giving serious thought to having the reservoir project beneficiaries carrying any cost of waste treatment in the upper areas beyond which that area would have to bear in the absence of the proposed reservoir.

2. Legal Action

The absence of any legal actions having been taken by the Commission is seen more as an administrative choice than a weakness of the basin agency mechanism. As has been stated on other matters, the Commissioners who have served to date have wished to exhaust all opportunities for local action before interceding. This "senatorial courtesy" to home rule has resulted in delays in obtaining pollution abatement, but it has also diminished fears of the possibility of preemptive action by the Commission in the joint exercise of sovereign power principle. Nevertheless, the Commission could, by simple majority vote, take legal action against one of the parties signatory to the Compact if the situation should so demand. There can be no doubt but that the possibility of Commission legal action has been a very important consideration in its ability to obtain its will without resorting to legal confrontations.

Two final aspects of the Compact warrant brief mention. These are the single vote per signatory agency and the budgetary support for Commission activities.

The degree of participation in basin matters, whether measured by land area, percentage of the signatories' drainage area lying within the Delaware River Basin, or monetary support, will not provide a rational basis for the one vote per signatory granted by the Compact; the

186

voting privileges were arrived at through the political process. After ten years of experience, the one vote per agency has proven to be highly successful. The Commissioners have looked upon the basin as a political entity unto itself and have voted as statesmen from that point of view.

The allocation of the annual budget among the signatory parties (for support of the Commission's staff) requires a unanimous vote. The work programme to be pursued during the budget-year is closely examined and redirected as deemed appropriate by the Commission prior to the overall cost of the programme and its division among the parties being approved. As drawn, the Compact would not relieve the Commission of any of its mandated basin-wide responsibilities if one or more of the signatory parties should fail to provide financial support. This contingency has neither been encountered or threatened in the ten-year life of the Delaware River Basin Commission.

CONCLUSIONS

The Delaware River Basin Compact contains a number of principles that could have application in implementing policies to avoid transfrontier pollution. Foremost among these principles are: (1) the concept of joint exercise of powers of sovereignty in the common interests of all the people of the basin; (2) the allocation of the water of the basin, or of the waste assimilative capacity thereof, in accordance with the doctrine of equitable apportionment; (3) adoption of a basin-wide, legally binding, comprehensive plan of water management; and (4) access to courts of competent jurisdiction to enforce duly promulgated decisions.

The fundamental strength of the Compact lies in its creation of a representative forum empowered to act, supported by legally enforceable authority in courts of competent jurisdiction.

Annex

ALLOCATION OF STREAM CAPACITY
FOR ACCEPTANCE OF WASTE DISCHARGES [1]

The water quality advisory committee consists of John Bryson (Delaware), Ernest Segesser (New Jersey), Donald Stevens (New York), Walter Lyon (Pennsylvania) and Edward Geismar (EPA).

The DRBC water quality staff participants are Ralph Porges, Head, Seymour P. Gross, Water Resources Engineer and David P. Pollison, Water Resources Engineer - Secretary WQAC.

1. This paper was prepared in consultation with the Water Quality Advisory Committee and reflects concurrence of the participating state and federal agencies. It was developed to provide further understanding of the allocation procedures as thus far developed.

Ralph Porges
Head, Water Quality Branch
July 23, 1971.

ALLOCATION OF STREAM CAPACITY
FOR ACCEPTANCE OF WASTE DISCHARGES

Background

The concept of waste load allocation is inherent in the Delaware
River Basin Compact, (1) as well as in the laws governing the functioning of various state agencies. The application of the doctrine of equitable
apportionment to water resources requires that the Delaware River Basin
Commission provide a means for sharing among the various water users
their equitable portion of the benefits to be derived from streams of the
Basin. One of these benefits is the assimilative capacity of streams for
the conveyance and ultimate disposal of treated wastes. This paper deals
with the allocation of stream capacity for acceptance of waste discharges
as related to dissolved oxygen in Basin streams.

The waste assimilative capacity of flowing waters can be scientific-
ally determined within reasonable limits. It, therefore, can be established
how much waste can be discharged so that the accepting water course may
still be suitable for desired uses prescribed by appropriate Standards (2)
and Regulations (3). This determination as it pertains to oxygen-demand-
ing wastes has been made for the Delaware Estuary, the 85-mile stretch
between Trenton, New Jersey, and Liston Point, Delaware, hereafter
referred to as the Estuary. This paper is limited to considerations of
allocations of the carbonaceous oxygen demand, also termed the first
stage biochemical oxygen demand (FSOD). As such, allocations of FSOD
will be in terms of permissible pounds per day, as prescribed in Section 3-3.11(2)a.(iii) of the Basin Regulations - Water Quality(3).

Allocation Basis

There are several basic premises underlying the allocations.
Fundamentally, these focus upon the objectives of the water management
programme, including the protection of existing and future water uses.
The establishment of appropriate criteria is required to assure adequate
water quality for the indicated uses. Throughout the Delaware River
Basin a minimum reduction of 85% biochemical oxygen demand (BOD)
is required. Where this treatment level is insufficient to sustain the
prescribed water uses or stream quality criteria, or to preserve existing high quality water, the stream's waste assimilative capacity can be
allocated among the waste dischargers.

Allocations are utilized where the minimum treatment requirements
will not protect existing and future uses in the receiving waters; therefore,
it is implicit in the water quality programme that such allocations may not

191

violate water quality criteria. Thus, the treatment level to meet an allocation may exceed but be not less than the basic treatment requirement of 85% of BOD removal. The effluent must also meet all other requirements set for bacterial concentrations, color, solids, toxicity, and so forth.

Development of Oxygen Demand Allocations for the Delaware River Estuary

In the Estuary, which borders on three states, the equitable apportionment principle was deemed applicable to the individual discharger irrespective of political boundaries, permitting consideration of the confined Estuary as an integrated system. For management purposes and in recognition of varying water uses and required water quality along the Estuary, it was divided into four zones (Figure 1). The equitable apportionment principle was implemented by requiring a uniform degree of treatment within each zone as of the date of the allocations. Any discharger may be required to increase treatment to offset an increase in its raw waste load in order to stay within its poundage allocation. Although zone boundaries are not inviolate, they do serve to exert certain restraints and to guide the application of the allocation.

Extensive study (4) of the Estuary was undertaken initially by the Public Health Service in co-operation with the DRBC, Basin state, local communities, industries, and other groups. This study developed a mathematical model which permitted determination of loadings that could be assimilated by the Estuary without violation of the stream quality criteria. These loadings, in terms of carbonaceous oxygen demand, are set forth in the Commission's Basin Regulations(3) and are as follows:

Zone 2	18,600 lbs. per day
Zone 3	144,800 lbs. per day
Zone 4	91,000 lbs. per day
Zone 5	67,600 lbs. per day
Total	322,000 lbs. per day

These loads are subject to review and adjustment by the Commission as increased knowledge and experience dictates.

The Regulations further state that as a part of the initial allocation and each subsequent reallocation, a reserve of about 10% of the total permissible load to the zone may be set aside for future needs. They further state that the pounds of carbonaceous oxygen demand minus the reserve will be allocated among individual dischargers; the allocations will be based upon the concept of uniform reduction of raw waste in a zone; each of the dischargers to a given zone will share in the assimilative capacity of that zone. The allocations for the base year 1964 resulted in zone treatment efficiencies as follows: Zone 2 - 88.5%; Zone 3 - 86%; Zone 4 - 89.25%, and Zone 5 - 87.5%.

Any treatment facility that was treating its waste beyond that required as a result of the allocation is required to continue that degree of treatment; the assigned allocation is based on this higher degree of treat-

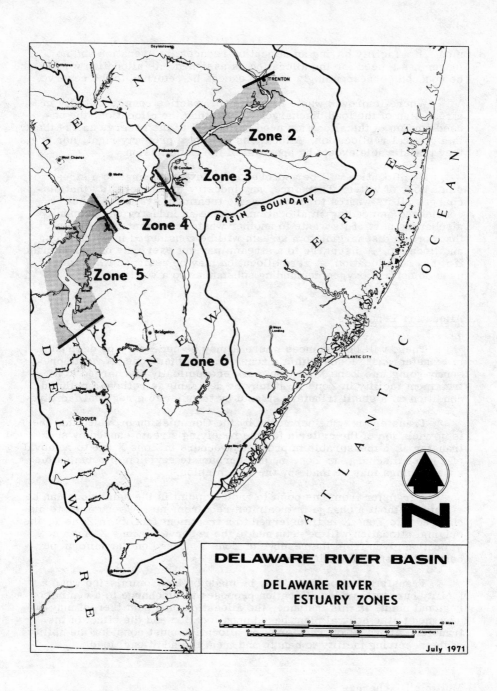

DELAWARE RIVER BASIN

DELAWARE RIVER
ESTUARY ZONES

10 0 10 20 30 40 Miles

10 0 10 20 30 40 50 Kilometers

July 1971

193

ment. If a facility having an allocation reduces its operation and no longer has a need for the allocation as assigned, its allocation would be reduced correspondingly and the excess be returned to the reserve.

Where improved waste management practice results initially in a reduction of the load discharged to less than the allocation, the unused portion of the allocation need not revert to the reserve unless there is a general reallocation. Allocations are not a property right, nor are they transferable except by approval of the Commission.

An allocation will be approved only for a discharge to a stream or other body of water. Therefore, any industry or municipality that discharges all its wastes to another waste treatment system will not be considered as needing an allocation. Where an industry or community discharges part of its waste to another waste treatment system, only that portion discharging to a stream will be considered in establishing an allocation. A discharge to a stream may not exert any oxygen demand if it has no allocation. A zero allocation may be assigned when water uses do not add oxygen demanding substances to a waste.

Allocation Transfers

There will be instances where it may be necessary or desirable to transfer waste loads within a zone or to another zone. A waste presently going into Zone X may be more economically discharged to a treatment facility in Zone Y. It may be desirable to relieve a polluted condition on a small tributary stream by transfer to a regional facility.

Transfers may be permitted by the Commission provided that there is no violation of the criteria for the receiving stream caused by such transfer. If a measurable improvement occurs in Zone X due to removal of a given discharge to Zone Y, an appropriate revision of permissible zone loadings may be made by the Commission.

A transfer from one point to another point in the same zone can be effected without a change in treatment requirements. Where a waste discharged into Zone X is transferred to a treatment facility in Zone Y, the original allocation will be returned to the reserve of Zone X and a new allocation given from the reserve of Zone Y, based on the uniform percent reduction applicable to Zone Y.

Transfer of allocations may be made by the Commission when a facility currently under allocation proposes to discharge to a central or regional plant. In this instance, the allocation or a part thereof may be assigned to the central plant based upon location and the effect of the transfer. Approval of transfer of an allocation must consider the ability of the receiving facility to handle and treat the additional load.

Multiple Discharges

The allocation to an industry is assigned to a specific facility and is addressed to the stream condition so that the entire waste from that facility will meet the allocation. Therefore, if an industry has several outfall sewers in reasonable proximity to each other at a given facility,

it is the industry's responsibility to control its discharges so as to
meet the allocation. It then is possible for the various waste streams
to be given different degrees of treatment to achieve the allocation
provided each of the waste effluents comply with all other requirements
set forth in state and DRBC regulations.

If an industrial plant or municipality diverts all its wastes to a
regional waste treatment system, their allocations may be transferred
by the Commission in their entirety or in part to the regional system.
There will be some instances where an industry may deem it advisable
to split its wastes in some fashion so that part will be discharged to a
regional waste treatment system and part, after treatment, discharged
directly to the stream. In this case, that part discharged to a regional
waste treatment system must be considered in the same light as any
other industry discharging to a regional facility and no allocation will
be assigned to said industry for this waste stream, and the allocation
transferred in part or in its entirety to the regional facility. The waste
stream or streams reaching the river directly would be assigned an
allocation based on the contribution of raw waste load prior to treatment.
In some instances the waste discharged to a regional facility may require
pre-treatment. This will be an arrangement between the industry and the
regional facility.

Future Growth

Industrial and municipal growth brings about increasing raw loads
which must be discharged after treatment to a stream with fixed assimilative capacity. Therefore, the waste treatment facility must be designed to increase the degree of treatment as the load increases. This is
inherent to the successful functioning of the allocation concept. Normal
growth must be compensated by upgrading treatment. Allocations for
new communities, new industries, industrial plant expansion, or community annexation may be made from the reserve based on the zone
degree of treatment initially employed for allocations and subsequently
established for reallocations.

Regionalization of Waste Treatment Systems

Regionalization is being encouraged by the Basin States and the
Commission as a means of expediting and economizing on waste treatment. A single large regional waste treatment system takes advantage
of economies of scale as well as reducing operation and maintenance
costs. There are many additional important advantages, such as the
buffering action and chemical neutralization of large volumes of waste,
and simplification of surveillance. The possible disadvantages of con-
centrating wastes at a single outfall must also be recognized and evaluated.

Should existing facilities under allocation join another plant or a
regional waste treatment system, allocations or part thereof may be
assigned to the other plant or to the regional system. New discharges
to a regional system not previously under allocation may necessitate an
allocation to the regional facility from the reserve, based upon the load
and the degree of treatment current in the zone at the time of hookup.

Normal growth of municipal or industrial contributions will be expected to be treated within the assigned allocation. Some regionalization proposals will involve trunk sewers and a treatment facility which must be built to serve a large service area, but which will initially serve a much smaller area. The initial allocation must then be based on the load generated within the initial service area and the allocation be subject to review as discharges from additional service areas are hooked up to the trunk sewer and the treatment plant.

In some instances it may be economically desirable to treat wastes that originate outside of the Delaware River Basin at a regional facility within the Basin. In such a case, the Commission has not recognized the imported raw waste load in the allocation. The added cost of such treatment will be subject to negotiation between the waste producer desiring treatment and the agency responsible for the regional treatment plant.

CONCLUSIONS

The assignment of allocations must be equitable. To assure success of this desirable procedure will necessitate the application of reasonableness and fairness to all concerned. The resources of the region are available to all, must be shared by all, and must be protected by all.

It is not possible to foresee all of the ramifications in the handling of allocations, but it is essential to recognize allocation as a means of maintaining our streams in a suitable condition. In this fashion, the allocation concept will encourage the establishment of industries and communities in the basin to further the economic and social well-being of the community.

REFERENCES

1) Delaware River Basin Compact, Delaware River Basin Commission, Trenton, New Jersey (November 1961).

2) Water Quality Standards, Delaware River Basin Commission, Trenton, New Jersey (April 1967), Section X, Comprehensive Plan.

3) Basin Regulations - Water Quality, Delaware River Basin Commission, Trenton, New Jersey (March 1968).

4) Delaware Estuary Comprehensive Study, Preliminary Report and Findings, Department of the Interior, Federal Water Pollution Control Administration, Philadelphia, Pennsylvania (July 1966).

APPLICATION OF THE SEPARABLE COSTS - REMAINING BENEFITS PRINCIPLES OF COST ALLOCATION TO A REGIONAL SEWERAGE SYSTEM

by

Herbert A. Howlett

Delaware River Basin Commission
Trenton, N. J. , U. S. A.

FORWARD

In 1969, the Delaware River Basin Commission initiated a study to determine the engineering, economic and financial feasibility of implementing a regional sewerage system for all or portions of a 600-square mile area in the State of New Jersey. While negative conclusions were eventually reached, several techniques developed during the course of the study may find usefulness elsewhere. Specifically, novel approaches were used in the following areas:

1. Allocation of capital costs among the participating parties.

2. Allocation of operating costs among the participating parties.

3. Operation of the proposed facilities by a not-for-profit organisation on behalf of the owner.

This paper discusses only the method used to allocate the capital costs of the proposed regional sewerage system among its users. It is a new application of a long standing technique - The Separable Costs - Remaining Benefits methods of cost allocation.

Acknowledgment is given to Engineering-Science, Inc., Washington, D.C., the Consulting Engineer for the Delaware River Basin Commission, from whose reports factual data have been drawn for use in this presentation.

INTRODUCTION

In the Delaware River Basin drainage area of the United States, the Delaware River Basin Commission [1] (DRBC) has been created by legislative action, and given the responsibility of managing the waters of the entire watershed. The DRBC has adopted a policy [2]

1) requiring the maximum feasible utilization of regional water pollution control facilities and

2) proclaiming that it may provide planning and, when necessary, constructing, financing and operating services required for regional solutions to water pollution problems in the absence of other appropriate agencies.

Under this general policy, the DRBC undertook an investigation of the relative merits of treating the municipal and industrial liquid wastes from a 600-square mile area in the State of New Jersey in a regional sewerage system. The study objectives were to:

1) establish the boundaries of a sewage collection area based on cost and other considerations,

2) develop criteria for the design, construction, and cost determination of a proposed regional combined industrial and municipal wastewater interceptor sewers, treatment plants, and sludge disposal systems,

3) develop a plan for equitable cost apportionment among the various industrial and public entities which may be served by the regional system, and

4) develop a financing programme assuring that the necessary funds would be available for construction of the proposed system.

Conventional studies of population, water use, waste quantities and qualities relating to both present and future conditions culminated in alternative physical plans. Present worth analysis identified an optimum plan and the task became one of testing that plan to determine

1) whether it was economically justified,

2) the logical boundaries of its service area, and

1. Delaware River Basin Compact, United States Public Law 87-328, Approved September 27, 1961, 75 Statutes at Large 688. Trenton, New Jersey, 1961.

2. Howlett, Herbert A. 1972. A Basin Agency and the Joint Exercise of Sovereign Powers, these proceedings (1972).

3) an equitable apportionment of its costs among the participating parties.

A review was made of accepted methods of cost allocations (see Annex) which disclosed that the separable costs-remaining benefits method of cost allocation, (SC-RB), contained principles which would be well suited to gaining insight to these issues.

Separable Costs - Remaining Benefits

The separable costs - remaining benefits (SC-RB) method of cost allocation has been widely used by federal agencies of the United States to determine the costs of a multiple-purpose project assignable to its individual purposes. While generally accepted by large agencies, it has not been as well received by other agencies due to the tedious mathematics needed to carry it to conclusion. As stated in "Proposed Practices for Economic Analysis of River Basin Projects"[1] - "The separable cost for each project purpose is the difference between the cost of the multiple-purpose project and the cost of the project with the purpose omitted." Thus, if the multiple-purpose project is being designed for flood control, power, irrigation, and navigation, it is necessary to develop a preliminary design and cost estimate for not only the total project but also for a series of projects which omits each of the purposes in sequence. Similarly tedious, the method requires computation of the benefits that will attain to each of the several purposes, as by definition, the method prohibits allocating more cost to a given purpose than the amount of its benefits.

Table 1.[1]   ALLOCATION OF COSTS
BY SEPARABLE COSTS-REMAINING BENEFITS METHOD
General Case   (In thousands of dollars)

ITEM	FLOOD CONTROL	POWER	IRRIGATION	NAVIGA-TION	TOTAL
1. Benefits	500	1,500	350	100	2,450
2. Alternative cost	400	1,000	600	80	2,080
3. Benefits limited by alternative cost (lesser of items 1 and 2)	400	1,000	350	350	1,830
4. Separable costs	380	600	150	50	1,180
5. Remaining benefits (items 3-4)	20	400	200	30	650
6. Allocated residual cost	18	360	180	27	585
7. Total allocation (items 4 + 6)	398	960	330	77	1,765

1.   Ibid.

The [SC-RB] method of cost allocation is illustrated above for a multiple-purpose project for which the total project costs amount to

1.   Proposed Practices for Economic Analysis of River Basin Projects, Inter-Agency Committee on Water Resources, Washington, D.C. May 1958.

$1,765,000. These include investment costs and operation, maintenance, and replacement costs, all reduced to a common time basis, and are expressed either as an average annual amount or a present worth amount.

DRBC Application of SC-RB Principles

The problems facing the DRBC were

1) allocating the capital costs of a single-purpose project among a number of participants in a regional sewerage system, some of whom would not utilize all of the features of the project, (the operating costs were apportioned utilizing a separate method), and

2) establishing the boundaries of the area to be served by the regional system.

The optimum plan included interceptors extending in opposite directions from a central treatment plant which, in turn, would be located adjacent to the largest contributor of waste. Thus, a customer on one interceptor would not make use of the other interceptor, and the customer adjacent to the treatment plant would use neither of the interceptors. Yet, it was rationalized, the existence of the interceptors would make possible the delivery of waste which, in turn, would result in economies of scale in the treatment facility - thus benefiting all. Accordingly, the total capital cost of the interceptors and treatment facilities was allocated among all participants.

The financial feasibility of the project depended upon the issuance of revenue bonds unsupported by general taxation authority. Therefore, it would be necessary to enter into enforceable contracts with the several participants for the repayment of their appropriate share of the overall capital cost. As each party was under legal orders to abate his pollution, he was faced with the option of either constructing his own facilities or participating in the regional system. As present worth studies indicated the economic superiority of the regional system, a technique was needed which would equitably apportion the project's capital cost among the participating entities.

The SC-RB method of cost allocation was modified to allocate the project capital costs to each participant and to provide insight into the economic justification and financial feasibility of the project. In addition, the modified SC-RB programme gave some indication as to the justifiable length of the interceptors. Lastly, as the problem lent itself to programming for modern computers, the tedious calculations heretofore encountered in the SC-RB technique were substantially reduced.

An illustration of the DRBC application of the SC-RB principle is shown in Table 2 for a fourteen customer project costing a total of $57,868,000. Note that capacity had been designed into the system in anticipation of future participants (FP). The total capital cost includes $26,813,000 for the interceptors and $31,055,000 for the treatment facilities.

In the general application of the SC-RB method (Line 1, Table 1), benefits must be determined for each project purpose. In lieu thereof,

203

Table 2.[1] CAPITAL APPORTIONMENT BY DRBC APPLICATION OF SC-RB PRINCIPLES

Capital Cost in Thousands of Dollars

ITEM	(A)	(B)	(C)	(D)	(E)	(F)	(G)	(H)	(J)	(K)	(L)	(M)	(N)	FP(c)	TOTAL
1. Justifiable Costs (a)	4,390	2,160	430	6,660	880	1,400	4,450	950	17,340	1,000	730	1,700	1,110	26,130	71,130
2. Separable Costs															
a. Interceptor	4,824	881	72	3,448	76	65	695	141	-	40	25	255	1,164	7,837	
b. Treatment Plant	1,486	582	48	2,767	31	56	621	252	9,504	242	172	475	290	4,057	
c. Total	6,310	1,463	120	6,215	107	121	1,316	393	9,504	282	197	730	1,454	11,894	40,806
3. Remaining Just. Costs 1 - 2c	-1,920	697	310	445	773	1,279	3,314	557	7,836	718	533	970	-344	14,236	30,504
4. Percent Distribution 3 ÷ 32,768	-	2.127	0.946	1.358	2.359	3.903	10.113	1.699	23.913	2.191	1.626	2.960	-	43.444	99.995
5. Remaining Joint Costs 4(57,863 - 38,542)	-	411	183	263	456	754	1,955	328	4,621	424	314	572	-	8,396	19,326
6. Total Apportionment 2C + 5	4,390	1,874	303	6,478	563	875	3,271	721	14,125	706	511	1,302	1,110	20,290	57,868(b)
7. Capital Apportionment (1 20,290 (Total Apportionment) 37,578) (Item 6 above)	6,761	2,886	467	9,976	867	1,346	5,037	1,111	21,751	1,087	787	2,005	1,709	20,290	57,868

1. Addendum to Final Report, Deepwater Regional Sewerage System Preliminary Engineering and Feasibility Study, Engineering-Science, Inc., Washington, D.C, December 1970.

in the DRBC application, purposes are replaced by participants and the cost of each participant constructing his own facilities is assumed to be his alternative costs, and the limit of his willingness to pay. Accordingly, "benefits" and "alternative costs" are assumed to be equal and were called "justifiable costs" in the DRBC application (Line 1, Table 2). The cost of each participant constructing his own independent facilities must be estimated using design and cost criteria comparable to that utilized for the regional system, i. e. , construction cost indices and degree of treatment should be the same. Methods of treatment can be different to reflect the most appropriate way of reaching the desired quality goal.

As in the general application of the SC-RB method, determination of separable costs in the DRBC approach (Line 2, Table 2) is accomplished by designing a regional facility, estimating its cost, then designing and costing a system, omitting a participant. The difference between these costs is equal to the "separable cost" of the participant in question. It should be noted that those who do not utilize an interceptor gain some credit in the procedure at this point. These calculations can be formulized and programmed for a computer, thus giving reasonable accurate estimates of cost, and eliminating the tedious and time consuming methods of the past.

Lines 2a, b, and c, Table 2, disclose that the separable costs for participants A and N are greater than their justifiable cost and point to the interceptor as the principal cause. This would indicate that the cost of including these two parties in the system exceeds the benefits of their participation. Further, as both were at the extreme ends of the two interceptors, it was taken as an indication that the interceptors were too long, and service area to large.

At this point, a decision must be made on whether to include participants A and N in the system. This can be looked upon as a political decision, i. e. , the other parties may desire to include their neighbours for reasons not relevant to this test. For purposes of illustration, it has been assumed that the decision was to include both parties. Therefore, the amount to be allocated in accordance with "remaining justifiable costs" equals the $30,504,000 shown on Line 3, Table 2, plus the amounts of participants A and N beyond their justifiable costs ($1,920,000 and $344,000 respectively) for a total of $32,768,000.

Line 4, Table 2, is the simple percentage distribution of the $32,768,000 among the parties of Line 3 taking A and N as zero.

With the political decision to include A and N in the system, and their participation limited by justifiable costs, the amount to be distributed in accordance with Line 4 is the total project cost, ($57,868,000) minus the sum of the separable cost, less the negative amounts of participants A and N ($40,806,000 - $1,920 - $344,) or $19,326,000. The distribution of this amount is shown on Line 5, Table 2, and the total apportionment on Line 6. Note on Line 6 that the apportionment of A and N is limited to "justifiable costs". Also note that an allocation has been made to future participants (F).

Future Participants. - The cost of providing excess capacity in a system to accommodate economic growth to an area must be borne by someone. In the present illustration, revenue bonds were to be the basis of financing, and enforceable contracts would be needed to underwrite the entire capital cost ($57,858,000). Therefore, the cost initially allocated to FP was distributed proportionately among the initial participants (Line 7, Table 2). In this computation, the limitations imposed by "justifiable costs" were not exercised. Note that this apportionment renders the project unattractive to most parties, on the basis of capital cost.

DISCUSSION

The foregoing presentation of the DRBC application of the SC-RB principles is an actual case whose purpose was to distribute equitably the cost of a regional system among its potential users in order to underwrite the issuance of revenue bonds. Initially, this method of cost apportionment was viewed with skepticism. Most parties had difficulty accepting a seemingly new and untested method of cost apportionment; most had to be persuaded that the logic of apportioning the costs of both interceptors and the treatment facilities among all parties was sound. Most of the alternatives suggested were based upon the proportional use of facilities principle, and all failed to distribute any portion of the interceptors to the large waste producer adjacent to the treatment plant.

Perhaps the bitterest pill for the initial users of the system to swallow was that of underwriting the cost of future participants. Their arguments on this point were largely overcome by counter arguments:

1) under the assumed plan, the project owner would be eligible to receive State and Federal grants which would help reduce the capital obligation of the initial users;

2) the regional system under public ownership would provide capital at lower interest rates than could be expected by the participants;

3) the regional system would provide substantial savings in operating costs over those that would result from individual systems; and

4) under the plan of repayment, as new customers "bought in" to the system, either the annual payments of the initial underwriters would be reduced, or the income from the new participants would be used to shorten the period of capital indebtedness.

The DRBC application requires, by definition, estimates of the cost of a participant constructing his own facilities (justifiable costs - Line 1, Table 2). As might be expected, each party wanted to make his own independent estimate, using his own criteria and unit prices. Also, most were reluctant to provide capacity for their own growth beyond a five-year horizon. These matters took considerable negotiations to bring the estimates to a comparable base with those of the regional system. However, after the novelty of the proposal wore off, the group accepted DRBC's application of the SC-RB principles for capital apportionment purposes. A major factor in this acceptance was the knowledge that a theoretically sound theory had simply been slightly modified to solve a practical problem.

The DRBC technique, when applied to several alternative systems for providing the same level of waste treatment to all parties within the study area, disclosed that the system that would have the smallest present worth over a 50-year period, would, on the basis of capital cost apportionment, include two parties who could build their own systems cheaper. As the present worth analysis included operating and maintenance costs, as well as capital recovery, the DRBC technique provides insight into the limits of the area to be served, but is not definitive, in itself, as to overall project economic justification. Decisions on economic justification, financial feasibility, and repayment must turn on many factors, only one of which is discussed herein. Nevertheless, the DRBC application of SC-RB principles of cost allocation proved to be a practical tool and was generally accepted as equitable by the parties at interest.

CONCLUSION

Proven techniques, designed for one purpose, may be modified to solve totally different problems.

—

Annex

PRINCIPLES OF COST ALLOCATION

by

J.W. Thursby

Delaware River Basin Commission

The use of one structure to provide more than one service allows the services to be provided at less cost than the total cost of separate structures for each service. The incremental cost of providing each function as an addition to the other functions of the combined structure should be less than the cost of the most economical single-purpose alternative means of producing similar benefits for that function. This being so, a basic principle of cost allocation is that the saving derived through use of the combined structure for numerous functions should be shared equitably by all functions. There are certain benefit-cost relationships which should be recognized, such as that no function be assigned costs in excess of its benefits or be supported by benefits attributable to another purpose, and no function be assigned costs greater than the cost of an alternative single-purpose project. The allocation must, of course, be consistent also with existing laws, treaties and compacts affecting the project plan.

There is general agreement that the following broad principles are applicable:

1) Each purpose should share equitably in the savings resulting from multiple-purpose construction, within the limits of maximum and minimum allocations.

2) The minimum allocation to each purpose is its specific or its separable cost, depending on the method of allocation.

3) The maximum allocation to each purpose is its benefits or alternative single-purpose cost, whichever is less.

4) Joint costs should be apportioned without regard to the ability of any particular purpose to repay its costs.

5) Legal priorities for the use of water and existing laws, treaties, and compacts must be recognized.

6) Cost and benefit estimates (including any alternative project used in procedure) must be genuine and reliable.

209

Various Methods of Cost Allocation

The Benefits Method. The benefits method allocates costs among the various project purposes in proportion to the value of the benefits produced by each purpose. In one variation, all costs are assumed to be joint costs and are allocated among the purposes in direct proportion to the value of their benefits. In another variation, "direct (specific) costs" are first identified with their respective purposes, and the remaining "costs of joint facilities" are allocated in proportion to benefit.

The Separate Projects Method. In the separate projects method, costs are allocated in proportion to what the costs of obtaining equivalent benefits would have been if separate, single-purpose projects had been built to serve each purpose. In the first of three variations, all costs are regarded as joint costs and allocated accordingly. In the second, direct costs are first identified and the remaining costs of joint facilities are allocated in proportion to the difference between the estimated cost of the alternative single-purpose project for each purpose and the direct costs of that purpose. The third variation is similar to the second, except that separable costs rather than direct costs are used.

The Alternative Justifiable Expenditure Method. In this method, the costs of joint facilities are allocated in direct proportion to the "remaining alternative justified investment" for each project purpose. This investment amount is defined as the smaller of either (1) the cost of the most economical alternative single-purpose project which will produce equivalent benefits - less any direct costs, or (2) the total value of benefits ascribed to the purpose - less any direct costs.

The Vendibility Method. This method allocates costs in proportion to the market prices of the project commodities or services. To the extent that the market price can be considered equivalent to per-unit benefits, the method and its variations are similar to those methods which use benefits as the allocation.

The Use of Facilities Method. This method is based on the concept that the cost of joint facilities should be allocated among the various purposes in proportion to their respective "use" of those facilities. "Use" is measured either in terms of the storage capacity provided for the purpose, or in terms of the quantity of water flow, or both.

The Priority of Use Method. The premise of this method is that the various purposes compete with each other to some extent for the use of water flow capacity or storage space. Some purposes are regarded as having priority over others, and the method is designed to give special attention to those priorities. The method identifies direct costs with their respective purposes and allocates the costs of joint facilities in a descending order of priority. The purpose with the highest priority is assigned only its direct costs plus a share of the costs of joint facilities equal to the lesser of either (1) the benefits less direct costs or (2) the cost of the most economic alternative project less the direct costs of the purpose.

The Incremental Method. This method identifies separable costs with their respective purposes and allocates all joint costs to that single purpose which is considered the primary function of the project.

The Direct Costs Method. This method is a variation of the incremental method, except that direct costs rather than separable costs are employed, and the costs of joint facilities, rather than joint costs, are allocated to the primary project purpose.

The Equal Apportionment Method. In one variation, separable costs are identified, and in the other variation direct costs are identified. Depending on the variation used, either joint costs or the costs of joint facilities are apportioned equally among the principal purposes of the project.

The Separable Costs-Remaining Benefits Method. The separable costs remaining benefits method (referred to as the SC-RB method) is a modification of the alternative justifiable expenditure method, differing primarily from the latter by identifying separable, rather than direct (specific), costs. Its objective is to allocate costs in such a manner that (1) each project purpose will at least be allocated its separable costs; (2) the total costs allocated to any purpose as a basis for repayment will not exceed either the benefits of the purpose or the costs of providing the same benefits by the most economic alternative project; and (3) within these maximum and minimum limits, a proportional sharing of the savings from the multi-purpose development will result.

When used as a basis for repayment arrangements, this method cannot result in an allocation of anticipated cost to a particular purpose that is greater than the benefits expected to be derived from that purpose or that is less than the expected separable cost of that purpose. In the application of the method, the separable cost of each purpose is determined by computing the cost of the total project with the purpose included and, again, with the purpose omitted. The difference in these total cost figures is the separable cost attributable to that particular purpose. After all of the separable costs have been determined, joint costs are allocated among the various project purposes in direct proportion to the "remaining benefits" ascribed to each purpose.

ESTABLISHMENT
OF INTERNATIONAL ENVIRONMENTAL STANDARDS -
SOME ECONOMIC AND RELATED ASPECTS

by

John H. Cumberland

University of Maryland, U.S.A.

CONTENTS

Contents

I. INTRODUCTION: THE OBJECTIVES

While the achievement of a reasonable balance between environmental and developmental objectives is one of the most difficult and challenging problems currently faced within most nations, it is not surprising to find that even more difficult problems emerge in the international management of shared environmental resources. However, despite the difficulties involved, the importance of finding effective methods for international environmental management is difficult to overestimate since it will ultimately be a major determinant of the quality of human life on earth. This paper will therefore examine some of the central economic and related aspects of the problem and suggest methods for establishing and administering standards for international environmental management designed to maximize long-run international welfare. The central hypothesis of the paper is that the problem of international management of environmental resources is sufficiently complex as to require the development of new knowledge and the design of new institutions for the effective use of that information, but that the difficulty of achieving these goals is not beyond the capabilities of human ingenuity.

II. CONCEPTUAL AND OPERATIONAL PROBLEMS IN INTERNATIONAL ENVIRONMENTAL MANAGEMENT

An essential first step towards the design of international environmental standards is an exploration of the major problems involved in achieving optimal balance between environmental and developmental objectives. Some of these issues have been explored by the author in a previous study of the role of uniform standards in international environmental management, and may be summarized here briefly as background to further discussion of methods for establishing international environmental standards. [1]

The general theoretical conclusion derived from environmental economics is that there exists an optimal target point for the use of environmental resources for assimilating wastes, and that this optimal level of environmental use also is the optimal target for pollution abatement. Furthermore, this point of optimality can be precisely defined

---

1. John H. Cumberland, "The Role of Uniform Standards in International Environmental Management" in Problems of Environmental Economics, Record of the Seminar held at the OECD (Summer 1971) pp. 227-277, OECD, Paris, 1972.

in theory as that point at which the marginal benefits of emission or use equal the marginal costs of treatment or abatement.

This principle is illustrated in Figure 1. With no control measures, producers and consumers tend to treat air, water, and other common property resources as free goods up to the point where their further use confers zero marginal benefits. This situation is shown by the point Y on the marginal abatement cost curve BB' in Figure 1. Producers emit a maximum amount of residuals, Y to the environment, since they are under competitive pressure to avoid incurring abatement costs so long as the use of the environment for disposal purposes appears to be cost-less, and their competitors are similarly avoiding incurring unnecessary pollution abatement costs. [1]

Figure 1

DERIVATION OF OPTIMAL LEVELS OF PRICE
AND USE OF ENVIRONMENTAL RESOURCES

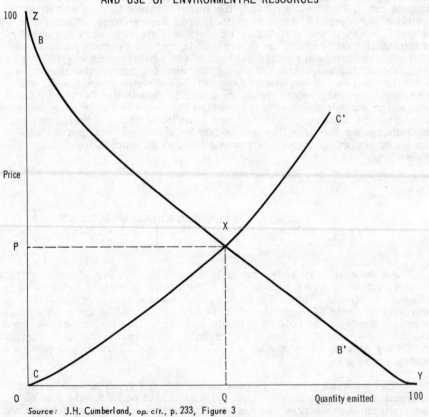

Source: J.H. Cumberland, op. cit., p. 233, Figure 3

However, the obvious wastefulness and inefficiency of excess pollution does not imply that achieving zero pollution should be sought or would be an optimal objective. For example, Figure 1 shows that the cost of achieving zero emissions (i.e. price Z at point 0) would be very

1.   Competitors may be activities within the same country or competitors in other nations.

high, and would not be warranted by the cost of the damage at low levels of pollution. In fact, permitting the use of the environment to absorb emissions up to point Q is less costly to society as a whole than attempting to reduce low levels of pollution below point Q, because abatement costs rise steeply (along BB') when efforts are made to reduce pollution further at already low levels of pollution.[1]

Therefore, given the functions shown in Figure 1, an optimal use by society of the environment for receiving residuals is point Q which can be defined as the point of Pareto optimality, because moving towards this point from any location away from it can be shown to make all parties potentially better off with none left worse off, whereas no movement away from point Q can be made without damaging the position of some party.

Although these concepts are widely understood and accepted in theory, the problems of giving the concepts quantitative and operational meaning, and of finding instruments for moving towards optimality in the real world are formidable enough even within sovereign nations and are vastly more complicated in the case of achieving agreements between nations.

In most cases, since the availability of common property resources encourages emittors to discharge wastes into the environment without counting the costs of the external diseconomies created, the problem is to reduce pollution levels to the optimal level. However, excessively restrictive regulations and control of less damaging emissions can conceivably result in the opposite error of requiring treatment activities which are more costly than the benefits achieved.

Therefore, the task of setting the optimal environmental standard is of paramount importance in environmental management. In theory, setting the standard can be easily achieved under certain conditions. If all parties have perfect knowledge of the damage functions and treatment cost functions, and if the parties involved own property rights to the environmental resource, and if the parties are free to bargain, the bargaining process should result automatically in an equilibrium situation at the optimal point. This identical solution should be reached whether the emittor is charged for the right to emit, or whether the receptor must pay the emittor to reduce his emissions, although the distributional effects will obviously be different in the two cases. Also, if the emittor must pay, he may have more long-run incentives for developing more pollution-free technologies. Conceptually, if the receptor had to pay for pollution abatement he too would have incentives to develop pollution-reducing technologies, though his ability to do so would presumably be less.

In addition to regulating the amount of emissions, the optimal point also has the characteristic that it determines the appropriate price for the use of the resource. In Figure 1 this equilibrating price is at P and will be reached under the assumptions stated, regardless of whether the environmental rights are assigned to the receptors, so that the emittor must pay for emitting, or whether environmental rights are assigned to the emittor, so that the receptors must pay for the abatement process.

1. Near zero or mere trace levels of emission may be optimal for some types of pollutants, such as radioactivity and other irreversible poisons, for which the marginal damage curve CC' would lie above the marginal abatement curve throughout the range of each including the lefthand origin.

In practice, most environmental problems arise precisely because the emittors and receptors do not have perfect knowledge or clear property rights to the environmental resource: a price is not paid and the resource is over-used by the emittor discharging wastes into common property resources, or the resource is thought to be under-used because of excessively restrictive regulations. Under these circumstances which are characteristic of most of the environmental problems in the real world, the task for society is to find some substitute for the automatic equilibrating process which occurs under bargaining conditions. Theoretically, this can be achieved under certain circumstances by controlling emissions and limiting them to the optimal amount Q if their quantity is known, or by setting a charge for emissions at the price of P per unit of emissions if this price is known.

In the case where perfect knowledge is available concerning the treatment cost functions and the damage functions but clearly defined property rights do not exist, it is still possible to achieve the optimality result provided only that a central authority is given the responsibility for regulating the resource, and that the authority decides which party is to bear the costs, or how they are to be shared. The central authority can then assure the optimality solution either by setting price P for the use of the resource and charging the emittor this price per unit of emission, or by charging the receptor for the cost of treatment, or by setting the allowable quantity of emission at Q and by letting the costs fall upon the emittor or the receptor or by arranging a compromise solution.

In the case of the two countries where the upstream-downstream problem exists, an additional constraint may exist under the extreme condition in which the upstream emittor is unwilling to participate voluntarily in any resource management arrangement unless he is able to extract from it the maximum gain which his spatial location makes possible. This is the amount of P x Q represented in Figure 1 by the rectangle OPXQ. If no co-operative alternative can be found, then this solution at least should be minimally acceptable to all parties, since all are better off at this point than at any other, even if the injured downstream party must suffer the added burden of paying the upstream emittor for reducing his emissions. However, it is reasonable to assume that in most cases, both countries would be willing to discuss mutually acceptable solutions, since few nations are consistently in an upstream situation vis-à-vis all emissions, liquid and gas, with respect to all other nations at all times. Furthermore, growing recognition that the environment of the earth is essentially a single closed system for whose wise management ultimately all nations must share responsibility suggests the hope that such cases of upstream environmental aggression will be rare. However, the possibility exists than an intransigent upstream emitter may commit environmental agression, and that fact should be recognized. In the worst case, he can extract the full price, O P for reducing pollution back to point Q, provided that he knows the damage function of the receptor, and provided the cost function, BB' reflects the full costs to the emitter of giving up his advantages gained from emitting. Therefore, even this most difficult case of international pollution can be resolved at high cost to the downstream receptor but with benefits still exceeding the costs.

Thus far the theoretical and conceptual problems examined present no insuperable obstacle to the international management of environmental resources. However, it must be recognized that the theoretical constructions used here are abstractions from reality, and that effective environmental management of international resources will involve an extraordinary range of difficulties. In the most general case, unlike this example of two countries and one pollutant, there will be multiple emitting countries, multiple receptor nations, with multiple plants in each releasing multiple emissions. The major difficulties involved will be those of obtaining the data necessary for defining the damage and treatment cost functions, deciding upon the allocation of costs and benefits, and developing institutions and procedures necessary for implementing management decisions. Each of these problems will be examined in turn. Of the large amount of information needed for environmental management that which is least difficult to obtain is data on the cost functions for reducing pollution, as represented by curve BB' in Figure 1. Although data on treatment cost functions are not abundantly available in the proper form, there do not appear to be serious obstacles to the development of such information, since it can be derived from industry studies, from producers of treatment equipment, from engineering studies, or even possibly from chemical and physical formulae. The important feature of such data is that they must indicate the cost of achieving various levels of emission abatement, in the schedule form shown in Figure 1, rather than simply presenting fixed costs for a single level of treatment, as is often the case. This schedule formulation is important because it is generally assumed that low levels of treatment can be achieved at low unit costs, but that achieving higher levels of treatment encounters sharply rising marginal costs. Another important feature of marginal treatment cost functions is that they reflect the costs of the full range of alternative treatment processes and options available.

Unfortunately, data on damage functions are much more difficult to define and to measure than are data for treatment cost functions. The physical, meteorological, and biological processes involved are highly complex. As physical emissions of liquid, gas, or solid wastes are generated by emittors, they are released into the air, or water, or deposited upon the ground. Diffusion processes then occur, which may absorb, dilute, assimilate, transform, and transport the emission to the point at which it impacts upon the receptor, which can be a human being or some plant or animal species of concern to human beings. In view of both the concern for conservation and the lack of basic knowledge about long-run ecological inter-relationships between species, it cannot safely be assumed that any species is totally without significance for the human population. Finally, as the emission eventually impacts upon the receptor, its ultimate effect upon him depends, among other things, upon his age, sex, weight, health, metabolism, and a multitude of other factors.[1]

The complexity of this process creates severe difficulties in setting environmental standards, and raises innumerable questions about the definition of standards, measurement of damage, and above all, about the valuation, or assessment, of the damage. A simplified diagram of some of the relationships involved is shown in Figure 2. The only variable normally subject to control is the discharge of residuals, or emissions, shown on the lower vertical axis R. As emissions are transformed by the receiving medium they affect the ambient environmental quality of the receiving medium. The relationships between emissions and ambient environmental quality is probably not linear, but is complex and curve-

---

1. For further information see the reference to Michio Hashimoto in Cumberland, op. cit. p. 236.

Figure 2

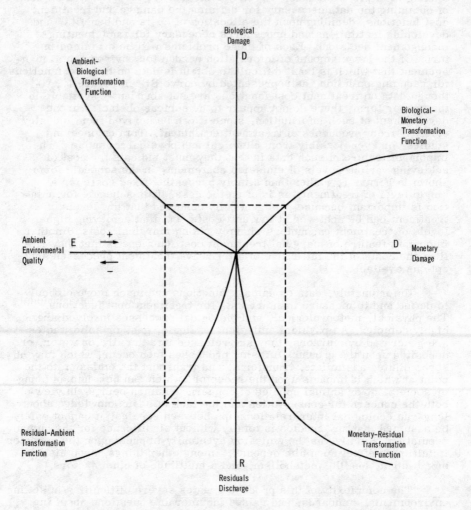

Source : Based upon A. Myrick Freeman III, "The Economics of Pollution Control and Environmental Quality" General Learning Press, New York, 1971, with suggestions for adding the Biological-Monetary Transformation Function by Henri Smets.

linear as indicated by the transformation function in the lower left quadrant of Figure 2. This relationship will obviously be affected by such
factors as local air currents, water flows, seasonal factors and other
variables and their interrelationships.

Once the shape and location of the transformation function is
known, the task of converting it into a biological damage function is
even more difficult. The impact of pollutants in the environmental
media upon living organisms is subject to all of the complexities mentioned by Hashimoto. Additional difficulties result from the well-known
problems of synergism, long-run accumulation, and related variables.
With few exceptions, scientists have not yet begun to structure their
investigations and findings in the quantitative, schedule type of form
shown in the upper left quadrant of Figure 2, which relates ambient
presence of pollutants to quantitative measures of health, longevity,
and well-being. Even less success has been achieved in evaluating
quantitative physiological damage functions in terms of monetary damage functions shown in the upper right quadrant required ideally in
Figure 2 for establishing environmental price and quantity standards.

Finally, in order to relate biological damage functions to various
levels of emissions (usually the major variable subject to policy control) the transformation function in the lower right quadrant is ideally
required. This is the ultimate damage function which is usually assumed to be necessary in order to equate marginal damage and treatment
cost functions for determination of optimal environmental emission
standards.

The conclusion drawn from this brief overview is that numerous
significant problems yet remain to be solved in the field of international
environmental management, particularly in relation to equity, policy,
basic information, and institutional development; these problems are
sufficiently acute as to warrant the consideration of new types of social
institutions designed to meet the problems. A proposal for a new institutional arrangement designed for international environment management
is considered in the next section.

III.   THE INTERNATIONAL INTERDISCIPLINARY COMMISSION
       FOR ENVIRONMENTAL MANAGEMENT: A PROPOSAL

A.   Introduction

The principal conclusion following from this examination of international environmental problems is that their solution will require both
the accumulation of a vast amount of information which is not now available, and effective new institutions designed to use that information for
expert environmental management. In order to improve the opportunities
for maximizing international welfare, this paper proposes the establishment of international interdisciplinary commissions of experts. The
major assumption made is that levying emission taxes on residuals
released into the environment is one of the most efficient instruments
available for environmental management and that part of the proceeds
from emission taxes can be used to finance the activities of the commissions.

B.   Assumptions

The major reason for proposing the establishment of interdisciplinary commissions of experts for international environmental management is the fact that movement towards efficient management will require more effective efforts than have been forthcoming to date to perform the extremely complex and difficult task of developing damage functions and cost functions, as demonstrated above in the discussion relating to Figure 2. The task of relating emission data to environmental quality, transforming this information into damage functions for humans and related species and then attaching value estimates to the damage functions will be a particularly challenging problem which can hardly be achieved in any other way than by interdisciplinary teams. In the case of international environmental management, the argument for an expert commission approach is even more compelling.

The second central element of this proposal is primary reliance upon emission taxes or fees as the major but not necessarily exclusive instrument for environmental management. Although strong alternative cases can also be made for the use of legislative regulation and even subsidization of pollution abatement and, indeed, the commissions might even make some use of these measures, there are strong reasons for giving priority to the principle of emission taxes. The basic economic and philosophical reason for this is that the use of the environment is not free or costless in a crowded, technological society, and that all participants need to be made aware of the impact of their actions upon the environment, and must help to pay their share of the real costs of protecting and managing the earth's fragile ecosystem. Preceding sections indicated how properly designed emission taxes can lead automatically to the establishment of optimal standards for management of residuals. Reliance upon emission taxes also has the advantage that, accepted as a basic principle and uniformly administered, they guarantee automatically a reasonable, if not perfect, international sharing of environmental management costs and they would solve many of the more difficult problems of how to allocate costs and benefits. Also, in terms of operational practice, more funds are needed immediately for environmental research and management purposes, and emission taxes can provide those funds. [1]

C.   Tasks of the Commissions

The commissions, which would be formed voluntarily at the recommendation of international agencies, such as the OECD, should include

---

1. Experimentation with interdisciplinary advisory commissions supported by environmental taxes used for research and administration has already begun in the State of Maryland in the United States. Funds are accumulated under the Power Plant Siting Act, which provides a variable surcharge on the consumers' monthly electricity bill. Money is used to support several study groups charged with making recommendations on environmental relationships between power plants and human health and welfare, monitoring emissions, commissioning research, siting power plants, and on related matters. Funds can be used to acquire the least damaging sites for power plants. This very promising beginning could be made even more responsive to optimality considerations by changing the financial arrangement from an ad valorem surcharge on the tax Bill to an actual unit emissions tax on radionuclides, thermal releases, particulate matter and other pollutants. At present these advisory groups are primarily limited in jurisdiction to a single state, rather than being inter-regional or international.

representation from all of the nations involved in the protection and management of each major water shed, air shed, and other environmental problem shed. Each nation would appoint delegates from among professional groups having sufficient objectivity, expertise and prestige, that their recommendations, ideally, would have the moral force of international legislation. The task of the commissions would be to improve the environmental performance of existing activities and, more basically, recognizing the dynamic cumulative nature of economic development, to design guidelines for future development which over time would assure environmental responsibility in balancing the demands of growth against the fragility of nature. The commissions would thus be corrective and future-oriented.

In support of these objectives, the specific tasks assigned to the commissions would be:

1) For each existing major source of pollution, use the best existing data, knowledge and judgement to develop cost functions for pollution abatement and damage functions for measuring benefits of changing levels of treatment.

2) Where existing data are not adequate undertake or commission research designed to provide the relevant data.

3) Using best available current and later new data on cost and damage functions, design and implement optimal tax structures for each emission at each location, supplemented if necessary with direct regulations, subsidy payments for treatment, and other measures where appropriate, including removal of the activities if in the public interest.

4) Using principles of objectivity, equity, fairness and good neighbour practices, recommend quantitative levels for international sharing of the contributions and expenditure of the proceeds.

5) For any proposed new construction or development of an activity which would emit residuals or which the commission decides would have a significant environmental impact, require the submission of an environmental impact statement [1] which would report fully on all expected residuals, accounting for how all inputs which do not become marketable outputs, are eventually discharged. The environmental impact statements should be based upon life cycle planning, covering fully the stages of exploration, construction, operation and eventual removal of the facility and restoration of the environment. The impact statement would be designed to provide all of the information necessary to permit the commission then to prepare a recommended schedule of emission taxes which would be applied at any specific location and level of activity proposed by the developers. Environmental impacts should be specifically defined to include aesthetic and amenity effects.

---

1. The submission of required environmental impact statements under the U.S. National Environmental Protection Act, has already become a major factor contributing to the flow of information, discussion analysis and management of environmental resources in the United States.

6) Beyond environmental analysis of existing and proposed activities within the jurisdiction of the commission, undertake the long-run design of a positive development programme, establishing goals and criteria for excellence in environmental management with specific concern for aesthetic effects, crowding, congestion and protection of areas for ecological, scientific, historic, and scenic purposes.

7) Where environmental emissions are not being adequately monitored, make provision for collecting all necessary data.

D. Some suggested Guidelines

High among the most potent forces for improved environmental management are discussion, education, fact-finding and public discussion. The commissions should encourage public hearings and publication of factual information on environmental matters in order to facilitate debate and to achieve support for their recommendations. In this spirit, it is especially important that the considerations behind the damage and treatment functions be subject to debate and review by experts so that both emittors and receptors can be given the right of challenge in an atmosphere of scientific standards for evaluation. Expert opinion is critical, but it should always be subject to public review. A sense of democratic participation is also important in making environmental decisions.

Initially, where management decisions must be made before the experts have all the final data they would like to have, they should be encouraged to present their best judgement based on available information concerning standards and taxes, [1] subject to review in the light of forthcoming data from research which the commissions should also specify. In difficult cases of uncertain information and critical need for environmental standards, sub-committees of experts, empowered to bring in additional consultants, should be asked to make interim judgements until adequate data are available; tentative standards and taxes can be applied, subject to observation of the results with options always existing for future adjustment. To protect the commissions' creditability and acceptance, taxes and standards should not be changed arbitrarily or capriciously, but should be changed when warranted by new or better information, or changed circumstances. Until adequate data are available, the commission should prefer to err on the side of caution. It should also attempt to avoid irreversible cumulative environmental events.

A central guideline for the commissions should be their future orientation based upon learning from past environmental errors and upon creating improved options for the future. One important way of doing this is to design their research so that it can shift treatment cost functions downward to the left, making improved environmental quality less costly through improved technology and exploration of other alter-

---

1. The part of this proposal which recommends setting an initial emissions charge, based upon the best available knowledge, but subject to subsequent revision, is consistent with the proposal, among others, by William J. Baumol and Wallace E. Oates in, "The use of Standards and Prices for Protection of the Environment", in the Swedish Journal of Economics, vol. 73, No. 1, March 1971, pp. 42-54.

natives. They should also consider shifting demand functions for environmental quality upward, to the right, where relevant, through research and education on environmental hazards.

As the work of the commissions advances, their management objectives should be to develop specific damage and treatment cost functions for each important emission, for each relevant actual or potential location. The serious problem of synergistic effects between residuals can be addressed by using waste matrices to account for all emissions resulting from specified activities, and by considering the use of synergysm surcharges over the standard emission taxes where potential deleterious effects can be anticipated. By the same logic, tax credits should be considered in cases of beneficial synergysm. In order properly to relate costs to damages, variable tax schedules should be considered when seasonal variations in temperatures, flows, biological cycles or other factors are important.

In all efforts to set standards, the commissions' work should be based to the extent possible on quantitative damage functions which indicate expected damages over different ranges of exposure of the organism. These quantitative damage functions should indicate the probabilities associated with specific health effects at different levels of exposure for different ages, sex and demographic groups.

In order to derive unambiguous price and quality standards, the quantitative treatment cost functions should be used in conjunction with damage functions in value terms, as in Figure 1. However, it may not be possible or even desirable to attach value figures to damage functions relating to human health. Nevertheless, quantification of the environmental impacts on human health, longevity and welfare for different demographic groups is the minimal information needed for informed establishment of environmental standards. If these data can be developed, emission tax rates and environmental standards can then be set on the basis of expert advice, informed judgement, political calculation and other bases. Then, after the results are judged in the light of experience, the emission tax rates can be adjusted, if desired, in accordance with any new preferred emission standard. In some cases, such as exposure to radioactivity, experts may wish to recommend only a single maximum exposure level, but they should be aware that operationally this implies a horizontal damage function of no damages up to the standard, and infinite damage beyond this point.

The relationship between the marginal damage functions and the marginal treatment cost functions will generate the appropriate emission tax levels and emission standards. However, as the author has pointed out in a related study mentioned earlier, the international differences in levels and composition of economic development, assimilative capacity and other factors will generate a variety of individual national and regional emission tax levels and environmental standards rather than uniform international standards. [1] Only in the case of biological concentration of pollutants in the human organism can a strong argument be developed for uniform international environmental standards, and even this case is weakened to the extent that cultural and other factors cause international variation in attitudes towards life, health, longevity, productivity and health. However, the very effort to achieve inter-

1. Cumberland, op. cit.

national co-operation and exchange of information on environmental factors may reduce international differences in attitudes towards environmental values.

In addition to its major tasks of collecting emission taxes and setting international environmental standards, the commissions should consider other appropriate actions. In exceptional cases they should consider using emission tax funds to compensate receptors who suffer from environmental impacts without other recourse to assistance.

The commissions should not necessarily be bound to use taxes collected from one emission source only for management of that emission, but should be free to apply judgement in maximizing general welfare through management of all environmental impacts within a total systems approach. Where warranted by the facts, the commissions should also be empowered to negotiate payments for the reduction of pollution or even to purchase activities for the purpose of closing them down in the case of serious damage without feasible alternative options.

If the commissions evolve as centres of expertise and prestige, they may be called upon to focus on or to participate in regional development, assisting in the identification of areas which should be protected from intensive development because they have long-run comparative advantages as areas of special scientific, ecological, scenic, recreational or other interest which are not fully valued by short-run market processes. Co-ordination between the commissions should also permit examination of opportunity costs and alternatives over broader ranges.

The proposed environmental impact statements which should be required to accompany the proposal for any new development having potential major environmental significance should be designed to include all of the data needed for a full-scale benefit-cost analysis. For the introduction of new technologies, a full technological assessment should be included, giving the commissions an opportunity to specify the desired operating standards and to guard against damaging side effects.

The pursuit of these objectives can be advanced by the adoption of a total systems approach which would attempt to examine the full range of direct and indirect effects from any proposed development. The benefit-cost approach should examine the full range of alternatives, including the elimination of the project and examination of the consequences of reduced rates of growth. [1]

IV.   CONCLUSIONS AND EVALUATION

This proposal that international organisations sponsor the creation of international interdisciplinary commissions for the development and administration of international environmental standards and emission

---

1.   This later alternative is particularly important for long-run considerations in view of some of the questions about exponential growth raised by the MIT Study, The Limits to Growth, Dennis Meadows et al, 1972.

taxes has not dealt with all of the important problems to be addressed, and it constitutes only one of many possible approaches which should be considered. However, it has numerous potential advantages.

The commission tax concept recognizes the right of the international community to a safe environment, and implements the principle that the polluter or the emittor should pay for his use of the earth's dwindling environmental resources. The proposal recognizes the existence both of international rights to the enjoyment of environmental resources, and the existence of corresponding international responsibility and opportunity for participating in management of the earth's fragile ecosystem. The emission taxes would raise revenue which is urgently needed for research to reduce serious gaps in fundamental knowledge about ecological and environmental relationships, in order to set international standards.

Emission taxes have the disadvantage of actually requiring the collection of money from emittors, but this in itself serves a purpose. Emittors would then take into account their impact on the environment, and would be presented with incentives for reducing their impacts. Any resulting price increases which were eventually passed along to the consumers of goods and services purchased from the polluting activity would provide further positive incentives for shifting resources away from polluting activities.

Finally, the rationale of the environmental commissions supported by emission taxes is future-oriented, recognizing that if significant improvements in technology and resource allocation can be built into the growth process itself, the international community could reverse the current trend towards cumulative environmental destruction in favour of improving the quality of life on the earth.

SOME TENTATIVE CONCLUSIONS FROM PROJECT EVALUATION EXPERIENCE WITH REGARD TO GUIDING PRINCIPLES ON TRANSFRONTIER POLLUTION

by

L. J. Locht

Institute for Land and Water Management Research,
Wageningen, Netherlands

It was agreed at the first session held by OECD consultants on transfrontier pollution (TFP) that this paper should contain specific suggestions based on earlier papers dealing with the problem in general. The paper is presented in the form of arguments followed by discussion. For a proper understanding of the subject, approaches to the general problem have been summarized beforehand.

I. THE PROBLEM IN GENERAL AND SOME DEFINITIONS

1.1   The theory illustrated by the graphs in fig. 1, which are taken from Muraro's paper, is typical of the economist's approach to TFP. The graphs illustrate a simple case of one-way pollution moving from an upstream country A to a downstream country B and imply some special assumptions regarding costs and damage. The diagram at the top shows the position of country A, while the diagrams in the second group show the positions of countries A and B, and of country B with abatement possibilities in B as well as in A. The diagram at the bottom shows the scope for bargaining between A and B.

1.2   In an economist's normal analysis, the institutional framework is a given datum, depending among other things on the rules for a Government's behaviour under international law. In the case of the present problem, however, the rules for Governments are the very thing which is under discussion, and in that context the economists present at the Seminar have tried to assess the consequences of possible alternative rules. The basic sets of rules and their consequences are taken mainly from the papers by Smets and Muraro and are given in the Table below. Definitions of the rules are given showing the resulting level of residual pollution and the resulting burden for each country calculated as the sum of its abatement costs and the residual damage it suffers.

For the argument in the paragraphs below it is essential to mention the following conclusions from this table:

a)   From the polluter's point of view, the order of preference is: MAP, VPP, PPP, FLP, while for the polluted country it is the other way round (with agreement);

b)   The polluter has no preference between VPP and no agreement, while for the polluted country no agreement is even worse than MAP;

c)   PPP as understood here (OECD definition, May 1972) means that "the polluter should bear the expenses of carrying out the measures decided by public authorities to ensure that the environment is in an acceptable state". As regards "acceptable state", the OECD report says that "In many circumstances the reduction of pollution beyond a certain level will not be practical or even necessary in view of the costs involved". This implies, as stated in the table, that if the standard is fixed at Q_g, coun-

try B is only partially protected against the effects of pollution by A and it still bears the burden MNLO as an externality. Only if the agreed standard is Q = S, being the standard that country

233

Figure 1

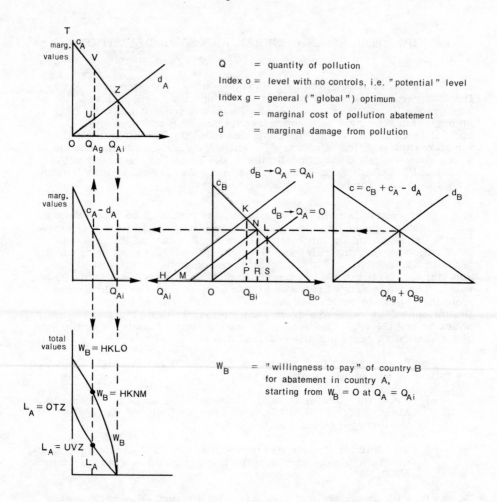

Q = quantity of pollution
Index o = level with no controls, i.e. "potential" level
Index g = general ("global") optimum
c = marginal cost of pollution abatement
d = marginal damage from pollution

W_B = "willingness to pay" of country B for abatement in country A, starting from $W_B = O$ at $Q_A = Q_{Ai}$

CONCLUSIONS :

a) The individual optimum abatement is $Q_{Ao} - Q_{Ai}$ in country A, $Q_{Bo} - Q_{Bi}$ in country B, and $(Q_{Bo} + Q_{Bg})$ for both together.

b) The "externality" in country B as a consequence of pollution in country A is represented by the area HKLO.

234

	COUNTRY A	COUNTRY B
I. Without agreement		
- residual pollution	Q_{Ai} (=OH)	OP + OH
- costs + residual damage	OZQ_{Ao}	OLQ_{Bo} + HKLO
II. Agreement with global optimum (Q_g) as standard		
- residual pollution	Q_{Ag} (=OM)	OR + OM
- costs + residual damage, with:		
. (PPP) Pollutor Pays Principle in OECD definition	OZQ_{Ao} + UVZ	OLQ_{Bo} + MNLO
. (FLP) Full Liability Principle for pollutor	OZQ_{Ao} +UVZ + MNLO	OLQ_{Bo}
. (VPP) Victim Pays Principle incorporating Fair Play Principle that A makes no profit	OZQ_{Ao}	OLQ_{Bo} + MNLO + UVZ
. (MAP) Market Allocation Principle incorporating that A makes a profit X, here some amount between W_B = HKNM and L_A = UVZ	OZQ_{Ao} - X	OLQ_{Bo} +MNLO+UVZ+X
III. Agreement without prefixed standard		
- residual pollution, with:		
. FLP and VPP	as under II	as under II
. PPP and MAP	O ⩽ Q ⩽ OH	OP ⩽ Q ⩽ OP + OH
- costs + residual damage, with:		
. FLP and VPP	as under II	as under II
. PPP (for Y see footnote)	OZQ_{Ao} + Y	OLQ_{Bo} + W_B - Y
. MAP with a profit X	OZQ_{Ao} + Y - X	OLQ_{Bo} + W_B - Y + X

Y is the cost of additional abatement which may be abatement in country A and thus lie on curve L_A, or abatement in country B and thus lie on c_B; in effect the lower of both curves.

B would have applied starting from Q_A = O, would the PPP do away with the entire externality. The costs for country A would be OZQ_{Ao} +UVZ+RNLS, the last term being a payment towards abatement costs in country B.

1.3 The papers by Bramsen and Stein state that from the lawyer's point of view, harmful pollution is forbidden, implicity referring to 'global' preference. A more specific rule would be, that one should not do to others what one does not want others to do to oneself. This rule refers to the preferences in country A. In essence these rules are clear enough. They involve, however, a jurisprudence which either must be based on FLP or PPP and, when applied to PPP, based either on Q_g as standard or $Q = S$.

1.4 There are some more remarks to be made with regard to terminology. Pollution is often used within a historical frame of reference, irrespective of whether it has a positive or negative utility. A case where pollution might have a positive utility would be a small increase in the normal CO_2 content of the atmosphere, which would increase plant growth.

Economists often do not include pollution with positive effects in their concept of pollution, in which case pollution has negative effects by definition. To avoid confusion it would seem preferable to speak of pollution with reference to some natural level, as biologists do. The economist's point of zero damage may therefore lie somewhere to the right of $Q = 0$.

Another consideration in defining pollution is the case of diversion of resources. This is not a form of pollution in the original sense of the word, but from the economist's point of view one would like to include it; e.g. in selecting a site for water storage for Rotterdam it made no difference whether the resource was expected to be polluted (Rhine) or whether it might be further diverted (Meuse). It seems confusing to include this kind of question under pollution and it would therefore seem wise to refer separately in the preamble to the guiding principles, to pollution and to diversion of resources.

In settlements between the Netherlands and Germany concerning minor rivers such as the Dinkel and Buurtsebeek, important points are the changes in peak levels of discharge. These cause, for example, changes in the locations of sand deposits and in the area flooded. This will probably not be regarded as a case of pollution, but it seems necessary to make the guiding principles cover such problems also.

In interpreting a presentation as given in Fig. 1, one has also to be aware of the difference between interfering with some particular goals, here called hindrance or harm, and damage. The hindrance is the original effect of the pollution, while damage is the effect taking into account the possibilities for adaptation of the individual and the economy. Examples of adaptation are departure from the polluted area, changes to the crop pattern, visits to alternative swimming facilities, etc. A special case is where it is not possible to abate pollution as such in country B, but where harm can be avoided. For instance, the Netherlands cannot abate the salt content of the Rhine, so that the presentation in Fig. 1 with a cost curve for B would not hold, but the Dutch can dilute, etc. the water from the Rhine (as discussed in section 2.1) and cope partly with the potential harm. This is illustrated in Fig. 2 and in this case a large part of the effects of TFP appears as costs (e.g. dilution costs) incurred by country B. Damage therefore is the residual harm as well as the cost of avoiding potential harm. The curve d_B, drawn from Fig. 1, is in fact the potential harm function, and the damage function is identical to the harm function up to the level L', while it also is an aggregate of dilution costs and the increase in harm from SL' to PK.

Figure 2

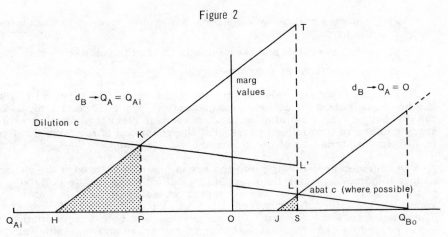

Illustration of the case where country B cannot abate TFP but can diminish the damage
(Note that we here refer to a case where some pollution (OJ) causes no damage).
The damage due to pollution is HKP, the potential damage due to pollution by A (externality) is HTLJ
and the real effect is HKP - JLS (for instance in agriculture)
+ PKL'S (for instance in water management)

II. SCOPE FOR DISAGREEMENT
ON THE EXTENT OF DAMAGE

My experience of evaluations of environmental factors is that the
estimates made are rough and reflect individual preferences which differ
from country to country preferences which sometimes change rapidly.
Neither third-country experts nor an international agency can obviate
that. The estimates leave wide scope for disagreement, even when extensive research has been done over the years. My contention is that the
guiding principles must be drawn explicitly with this fact in mind, that
is to say:

a)   guiding principles must not incorporate optimization models
which would make them unworkable; the alternative approach
is to formulate the guiding principles in terms of standards,
i. e. agreed levels of pollution;

b)   if a procedure could be formulated for making countries act
on their own estimate of damage, it might be a workable approach, so that the casual suggestion by Scott to introduce a
system of treadeable rights (pollution certificates) deserves
attention (see section 5).

2.1   An example of a damage estimate

For the information of the International Commission on Abatement
of Pollution of the River Rhine, the Section dealing with agriculture
presented a paper on the economic value of reducing the salt content of
the Rhine for Dutch agriculture, with special reference to horticulture.
This paper will be briefly discussed below to illustrate that conclusions
on such matters leave wide scope for disagreement, even in a case where

a) it seems relatively easy for a general economist to calculate the damage,

b) much research has been done in the field over many years,

c) international experts co-operated.

Starting from chloride load levels of 275 kg/s, 225 kg/s and 175 kg/s one had to calculate the chlorine concentration (in mg Cl per litre), using the discharge. The Commission took an average discharge over an arbitrary number of months and years, but the reactions of the farmers depend on the time distribution of the chlorine concentration.

Chlorides are entering into the areas concerned through the water supply to compensate for the evaporation deficit and to dilute polluted water (mainly caused by saline seepage from sea water). This water supply comes either directly from the Rhine or via the Yssel Lake, etc. The experts took the current situation as a basis for drawing their conclusions regarding damage, distinguishing three areas and assuming implicitly that there were no costs involved in creating and maintaining the current situation, presumably because they did not regard the existing measures as abating pollution. For those polders which are supplied with water via the Yssel Lake because of the Cl content of the Rhine in summer, however, one could include the opportunity cost of reserving storage capacity in the Yssel Lake, the cost of new supply canals and perhaps even part of the high costs incurred in recent years in diverting more water to the Yssel and less to the Lower Rhine, with consequent diversion of shipping and impacts on regional growth, etc. In actual fact, the pollution of the Rhine is not abated in the downstream country, but the harm is reduced and that at a price. With reference to Fig. 2, HKP has been calculated but not PKLS, which is a major part of the damage and the most difficult to assess.

In one of the areas selected by the experts, the water inlet point has been moved from near Rotterdam to near the town of Gouda. If the increment in TFP is additional to the former level of home-made Dutch pollution, there is sense in adding these costs to the effects of TFP, but if the increment in home-made Dutch pollution is added to some former level of TFP, one should not include it in the effects of TFP. One would have to make a shadow regional siting of the polluting activities in B and a related cost-benefit evaluation of the actual siting to arrive at the "proper" cost-sharing between TFP and home-made Dutch pollution. As a matter of fact, most of the current water management projects aim at more goals than just abating the effects of TFP and it seems practically impossible to disentangle objectively what part of the cost should be attributed to the TFP.

The chlorine concentration in the polder canal differs from the chlorine concentration on the water supplied to the area. It may suffice to mention that seepage, rainfall and evaporation (once again with their time distribution patterns), and the quantity of the water supply, have to be taken into account. The experts deduced that if there were 145 mg Cl/l in the Rhine, 230 mg Cl/l would occur in one of the areas, explicitly correcting this to 215 mg Cl/l to allow for bringing the water supply measures up to a "maximum". However, the cost of letting in more water and pumping it out again is part of the effects of TFP. Assuming "maximum" use of water resources, the experts obtained for the here selected region figures of 215, 192 and 170 mg Cl/litre from respectively 145, 118 and 92 mg Cl/l in the Rhine.

After taking into account other salts and the complex interrelation-ship between salts, soil and crops, the experts obtained for different soils and different crops continuous graphs for gross production and then went on to more technical problems. From the economist's point of view it is essential to take as a basis the existing distribution of horticulture between the different regions, as well as the existing types of crops and levels of gross production, but each of these factors has changed a lot since the Commission's report was written.

Another major point is that the situation in agriculture is far from equilibrium, the marginal productivity of labour, for instance, being below the wage level. When evaluating a rural project under these condi-tions, market values have to be corrected (by using a simulation pro-gramme; see section 3.1). With regional underemployment outside agri-culture (with full employment in the nation), it would also be necessary to trace the multiplier (with an input-output matrix).

Bearing in mind the many complications and the ad hoc decisions necessary to obtain quantitative results and the difference between residual harm and damage, it is understandable that the French and Dutch estimates of the damage ranged from about zero to a level warranting pollution abate-ment measures.

III. WHAT SHOULD BE COVERED IN CALCULATIONS OF THE ECONOMIC EFFECTS OF TFP

In preparing an approach to the evaluation of improvements in re-creational facilities, I have met with some fallacies in the economist's usual procedures and when trying to implement my conclusions I have encountered growing suspicion from researchers in other disciplines and also from the general public. Both the fallacies and the suspicion seem to be related to the general practice among economists of accepting income as a sufficient yardstick and doing so on the assumption that there is a reasonably functioning price mechanism. This has led ecologists to propose other value scales and guiding principles, which enjoy much popular support, and physical planners to propose their own evaluations. The resulting confusion encourages engineers to cling to their own (arbi-trary) standards.

This is relevant to the formulation of guiding principles for TFP, because

a) the computation procedure for damage is in principle the same as for improvements;

b) some calculation of damage will be the basis for any settlement of TFP claims, either under the formal "marginal costs equal to marginal damage" rule, or when agreeing on standards, or when claiming compensation in the courts for "serious" damage.

It seems therefore important to deal frankly in the preamble with the above-mentioned suspicions. The statement might run that TFP has effects, not only on production and consumption, but also on non-income goods, such as assurance of survival, and that these impacts have to be given due consideration in calculating damage. Another consequence is that it would be useful to devote, as a guide to economic research, a paragraph in the guiding principles to some of the essential points which calculation of TFP should cover. These points would include regional development (discussed in section 3.1), utility distribution (discussed in sections 3.2 and 3.3) and 'ecological' economics (section 3.4).

3.1   Regional development

Factor mobility seems often much smaller than assumed in conventional theory. Assessing benefit or damage on the assumption that there is approximate equilibrium, as is done in ordinary Cost-Benefit Analysis (CBA), therefore often amounts to ignoring the essence of the effects, which is a process of cumulative causation. Many economists are aware of this fallacy (e.g. Myrdal, 1957). Some discuss it as an objection to CBA as such (e.g. Fano, 1972); however, one can proceed within CBA by simulating what will happen in the course of time with and without the project. A simulation programme for this approach within the realm of agriculture was dealt with elsewhere (Locht, 1969).

This is relevant in the case discussed in 2.1; it is also for industrial pollution in Alsace. Economists will have to assess future developments in this area on the alternative assumption that the mines are closed down or that they are not. If they do not do this and draw conclusions which assume too much factor mobility, as in ordinary CBA, the estimated cost of closing the mines will be too low (probably even showing a profit).

3.2   Utility distribution

In a first version on a CBA study evaluating recreational facilities, I used the Clawson procedure starting with a quantity-distance relation and expressing distance in terms of price and steps were taken to estimate this relation for the old as well as the improved "product". Apart from that, shifts in the demand and supply curves over time were introduced. The relevant area enclosed by the demand curve (i.e. the willingness to pay) was assumed to represent the benefits yielded by the project. Where differences in travelling costs were not included on the costs side, the benefits were shown as a consumer surplus (Locht et al, 1970).

The essential alternatives in this case were:

a)   a public park

b)   space for building holiday dwellings.

The conclusion reached was that one should build a large number of holiday dwellings. To illustrate the fallacy of the economic argument here, it is easier to take the more simple comparison between (a) ficilities for horse-riding provided as a non-income good, and (b) facilities for fishing provided as a non-income good.

For a given utility (U), the willingness to pay (W) for improved horse-riding facilities will be high, because the income (Y) of the people concerned is high and $W = W(U, Y)$.

The calculation model therefore concludes that it is preferable to improve horse-riding facilities rather than fishing facilities, but this implies using tax revenue to provide more non-income goods for people with higher incomes. Even if one accepts an unequal distribution of incomes and with it an unequal distribution of income goods, this does not mean that one has to encourage an unequal distribution of non-income goods. Several economists have already pointed out this objection (e. g. Seckler, 1967; Poupardin, et al., 1972).

The case of second or holiday dwellings raises the same problem, because the consumer surplus on such dwellings involves changing the distribution of utility among individuals.

The argument may be summed up with the aid of Fig. 3. The utility level above a person's bare existence is represented by the area under the curve D_1 over the range Q_1° to Q_1^*, where Q_1° depends on his income, plus the area under the curve D_2 over the range Q_2° to Q_2^*, depending on the non-income goods available. I prefer to refer to that side as 'non-income goods, including free time. If a dimension is added to D_2, such as horse-riding facilities, the personal utility increases and the interpersonal utility distribution is changed. That is so because the user of the facilities does not have to pay for them. Only in the special case of a move by the government to increase the taxation of this group in proportion to the additional supply of non-income goods would it be justifiable to calculate on the basis of an uncorrected willingness to pay.

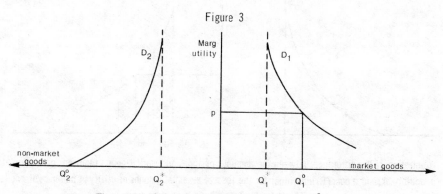

Figure 3

The utility for a person with income equal to $Q_1^\circ \cdot p$ is,
(apart from its bare existence, for which $Q_1^* + Q_2^*$ is needed)
the area enclosed by D_1 over the range Q_1° to Q_1^* plus the area enclosed by D_2 over the range Q_2° to Q_2^*

In a revised version of this CBA study, the derived W was corrected by using the reciprocal of the relative marginal rates of taxation as an approximate indicator of the collective assessment of the marginal utility of income. This may be correct as a rough estimate. Another argument for doing this is the implied consistency in a government's behaviour when it withdraws or supplies utility.

A second example will also be briefly discussed. In a recent study made in the Netherlands (Donnea, 1970), the willingness to pay for free time, which is a non-income good, was derived from the use of traffic modes and the author concluded that W for free time was higher in the upper income brackets than in the lower. I have objections to the deduced relation, but they are not relevant here. From the higher values placed on free time, it was argued that roads for people with higher incomes should be given a higher priority. But if a government behaved in that way and accepted increased traffic congestion where lower income groups are involved, more free time (a non-income good) would be withdrawn from those at a lower utility level, because they cannot pay so much for income goods (the traffic modes). Many more examples can be presented. Some often are very important within the context of our problem, e. g. the calculation of pollution damage in slums, where W will always be low. It therefore seems useful to devote a few lines in the guiding principles to countering the idea that a sound economic analysis would not include the effects on utility distribution.

3.3 Pareto compensation

Incorporating non-market goods, as discussed in section 3.2, in the usual formal optimization would involve paying monetary compensation, in terms of D_1 for effects expressed in terms of D_2. Taking the example of polluted swimming facilities for young people of limited means, the compensation would be as indicated for B in Fig. 4; this

Figure 4

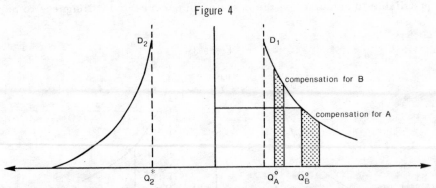

Illustration of the difference between the amounts of money and market goods required to compensate a wealthy (A) and a poor (B) individual for the loss of an equal amount of utility in non-market goods. The dotted columns have the same area.

means that compensating for an equal utility for rich and poor taken from the D_2 side implies providing a larger amount of income goods for those with higher incomes, as indicated for A in Fig. 4.

Damage studies calculating monetary equivalents in the ordinary way for changes in the D_2 goods, will lead to conclusions of this kind.

3.4 "Ecological economics"

Economists have discussed benefits and damage from the democratic point of view of the sovereignity of the consumer and have developed the science of utility economics, but now ecologists take another point of view based on a kind of recycling economics, of which there had

242

already been earlier examples. Just as they study animals and vegetation from the point of view of the suitability of their habitat, so they study man in relation to his real conditions of life and disapprove of over-consumption, the affluent society, etc.

For an economist it is easy to support their approach. The value of recreation becomes, instead of its utility, its effect on physical and mental health, measured perhaps in terms of productivity (Klaassen, 1972). The disutility of noise can no longer be assessed from behaviour with regard to the price of houses (as in Heath, 1969), but is valued according its effects on health. If a government acted on values calculated in the latter way, it would be taking a technocratic point of view. An extreme case of inconsistency between the two approaches would seem to be the evaluation of damage to fishing opportunities. The value of fishing in terms of direct utility, even after correction for the disutility suffered by the rest of the family, seems to be positive, but in ecological economics one may find it treated as often being an indulgence in day-dreaming whose effects on health and productivity are at best neutral. There are many similar cases, such as the disutility for fishermen of cormorants, which eat lots of fish, but have a positive value in recycling economics.

The lawyer's approach, where the reference to "harmful" is as cited in section 1.3, would seem to be close to that of the ecologist as regard health. In the last resort, however, the definition of health might not be very different from that of utility. In order to enlist general support for the guiding principles on TFP, it seems essential to specify explicitly that ecological values must be taken into account. To provide a basis for negotiations on TFP, it seems to this author that the two approaches have to be reconciled, perhaps with the aid of the old concepts of merit and demerit goods, but more probably along the lines to be discussed in section 4.

IV.   THE DOMINANT NON-INCOME GOOD INVOLVED

Utility economics have to be in accord with the psychologist's view on motives. Their list will start with:

- assurance of survival, being the need for a package of goods and services as represented in Fig. 3 by $Q_1{}^* + Q_2{}^*$,

- additional social contacts, self-fulfillment, etc.

Once these requirements are stated, the list should refer to:

- the unessential increments involved in the economist's marginal theory regarding a wealthy society. It is only these which are reduced when pollution abatement costs are incurred.

Therefore the damage (disutility) involved in changing the environment is extremely high when:

a)   one of the dimensions of $Q_2{}^*$ is affected, or

243

b) the level of Q_2* is not known, which is often the case, because the ecological relations are unknown, and consequently the future disutilities also, and

c) there are no opportunities for avoiding harm, so that potential damage equals potential harm.

In cases (a+c) or (b+c) it is certain that marginal costs will be below marginal damage and that abatement will be efficient from the point of view of international preferences. It will also be efficient from the standpoint of country A, if there is some positive "identification", as will be discussed in section 4.1. In this case the arguments about difficulties in quantifying damages are not relevant. It seems probable that, if the desire for assured survival is represented in D_2, one would find kinked

curves for the damage function, the kink being around the biologist's standard for those cases, so that it would be useful to state in the guiding principles that, from the economist's point of view as well, abatement beyond that kink is a minimum requirement. In some cases this will involve complete abatement (making the consequences of PPP equal to those of FLP). Some such organisation as the OECD could at a later stage provide a list of pollutants showing levels (in varying circumstances) at which survival may be threatened. Assuming the kinked curve, even in a zone concept, one would not require economists to calculate beyond the kink in order to arrive at maxima for the standards.

4.1 Identification

In dealing with TFP in ordinary economic analysis one assumes that the inhabitants of country A have no interest at all in any damage to country B's inhabitants and that people in country B have no interest in the costs caused in A. The lower graph in Fig. 1 is a clear example of this approach. The assumption, however, is not correct: there is a range of possibilities between being interested in

a) a utility for those in the other country as much as in the utility for one's own countrymen, and

b) a disutility for those in the other country.

In the Netherlands there is a historical background to the rules with regard to TFP. When there were small independent states, the changes occurring in the amounts of water discharged constituted damage over which wars were fought, but as the contacts between them grew closer, the differences between the people inside a "state" and the people in another one diminished, leading to a union.

My opinion is, therefore, that if one tries to build an analytical model for TFP, as is involved in the lower graph in Fig. 1, one has to include an "identification" factor (i) to indicate the degree of interest in the utility enjoyed by the other group. Pfaff and Pfaff (1970) refer to this element as the 'Boulding optimum' in contrast with the Pareto optimum.

Such an analytical model would yield wider scope for negotiation than that shown in the lower graph in Fig. 1 when i was negative, and a narrower scope when i was positive. The weight somebody would attach to the damage or cost to those in the other country would depend on i and also on the differences between the wealth of the two countries and the sizes of their populations. The argument may be summed up with the aid of the following formulae:

244

Let C and D be respectively the pollution abatement cost in country A and the damage in country B (both in \$), P_A and P_B the populations, and Y_A and Y_B the average incomes in these countries. I assess the approximate marginal utility of money (ψ) per \$ as

$$\psi_A = Y_g/Y_A \text{ in A and } \quad \psi_B = Y_g/Y_B \text{ in B.}$$

If an inhabitant of A with an average income were asked to vote on a special abatement project, his voting favour would depend on

$$\frac{C}{P_A} \cdot \psi_A + i^A \cdot (P_A - 1) \frac{C}{P_A} \cdot \psi_A < i^B \cdot D \cdot \psi_B$$

where:

i^A and i^B are the identification rates for country A's inhabitants with respect to their compatriots and with respect to B's inhabitants.

For a large country (a large P_A), this amounts to

$$\frac{C}{D} < \frac{i^B}{i^A} \cdot \frac{Y_A}{Y_B} \text{ and with } C' = \frac{C}{P_A} \text{ and } D' = \frac{D}{P_B} \text{ to}$$

$$\frac{C'}{D'} < \frac{i^B}{i^A} \cdot \frac{Y_A}{Y_B} \cdot \frac{P_B}{P_A} .$$

If an inhabitant of A voted on a representative instead of on a special project, he would add C over a range of projects and $\Sigma C/P_A$ would be substantial. Nevertheless, he would also in this case not be wholly indifferent to B's relative damages, relative income and relative population. The practical side of this contention is a proposal for an inquiry among representative groups in the OECD countries concerning their i rates and to include these i rates in the optimization models on TFP.

V. ADAPTATION OF PPP TO TRANSFRONTIER POLLUTION

The disutility (damage) depends on subjective preferences, so that for the same quantity of pollution it may vary from one area to another. This may be related to a "fear" that Q_2^* has been, or will be, reached.

An illustration of the different degrees of concern with the environment is provided by the sales of the Report of the Club of Rome, of which more copies were sold in the Netherlands than in all the rest of the world. However, standards enforced by international courts, or agreements under which the polluter pays a part (PPP) or the whole of the cost (FLP), will implicitly or explicitly be based on either, as stated in 1.3:

a)   average preferences in the international community, or

b)   average preferences in the polluting country.

Neither of these yardsticks results in optimum satisfaction of the polluted countries' preferences, so that in formulating the guiding principles one has to remember that there are two damage functions involved, one (h_B^g in Fig. 5) representing one of the rules a or b, and one representing damage as rated by national preferences in country B (h_B^B). I have assumed in Fig. 5 that the population of country B has a different estimate of Q_o^*, but that otherwise it shares the same preferences. However, the point also holds good if concern in regard to the environment is marked by other differences.

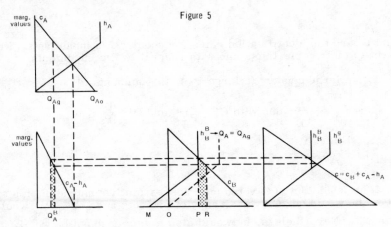

Figure 5

Illustration of a case where country B's inhabitants are more "afraid" of pollution, so that the kink in B's damage function (h_B^B) is lower than it would be if B has "normal" preferences (h_B^g). Additional abatement in A (the small column represents the cost involved) would be required to enable B to maximise utility as based on its own preferences.

Country B can claim abatement up to Q_{Ag} under PPP and FLP without reaching its own utility optimum. If PPP is given the same definition as it was by the OECD (1972) for domestic policy inside a country, so that all abatement costs have to be borne by the polluter, B could not then abate beyond the international standard and thus could not reach utility maximization as based on B's own preferences. With reference to Fig. 5, B could not, under the definition of PPP, abate from OR to OP, either at home or in A. Thus, in dealing with transfrontier damage, it seems essential to consider international preferences and to give the polluted country the right to require more abatement under VPP. This argument can be regarded as justifying the Netherlands' proposal to pay part of the abatement costs in Alsace without dropping PPP and without setting a precedent for sharing the costs of abatement up to international standards. A reasonable procedure would be as follows:

a) the polluting country would abate (PPP) up to the standard agreed upon as reflecting international preferences;

246

b) the polluting country would present its net cost function as the supply function for further abatement (internationally controlled);

c) the polluted country, if it had additional requirements would pay the cost (VPP) of the additional abatement.

The suggestion by Scott that one should introduce tradeable "rights" to pollute (pollution certificates) seems to me a good alternative to b and c, because it has the operational advantage that one need only provide "objective" standards and not cost functions as well. One must, however, realise that over this range of abatement it applies the market principle instead of VPP, but within the OECD group of countries, in which I assume that the identification factor is positive, this can be made acceptable.

In fact there is a serious obstacle to arriving at any agreement at all on TFP, if the polluter-pays principle has to be applied, because:

a) both parties can only with difficulty (publicly) accept standards involving high levels of residual damage, and

b) the polluting country will be opposed to low standards under PPP because of the cost involved.

To arrive at any agreement at all under PPP, the two-standards approach developed in this paragraph for such kinds of pollution as the salt load seems essential.

247

REFERENCES

BRAMSEN, C. Bo, 1972, this book, p. 257.

DONNEA, F. X. , 1970, Economische waarde van reistijdbesparingen in Nederland, Rapport Stichting het Nederlands Economisch Instituut, Rotterdam.

FANO, P. L. , 1972, Evaluating alternative plans configurations. Paper Reg. Sci. Ass. Conf. Rotterdam.

HEATH, J. B. , 1969. Cost-Benefit Analysis and Airport Location, in "Cost-Benefit Analysis", Ed. M. G. Kendall, The Engl. Univ. Press, 1971, London.

KLAASSEN, L. H. , 1968. Social amenities in area of economic growth, OECD, Paris.

LOCHT, L. J. , 1969. Evaluation of rural reconstruction, in "Cost-Benefit Analysis. " Ed. M. G. Kendall, The Engl. Univ. Press, 1971, London.

    , H. J. PROPER en G. HOOGENDOORN, 1970. Economische beoordeling van voorzieningen voor recreatie en natuur in Midden-Maasland, Nota ICW 623, Wageningen. (Revised version, 1971.)

MURARO, G. , 1972, this book, p. 75.

MYRDAL, G. , 1957. Economic theory and underdeveloped regions, Gerald Duckworth, London.

OECD, 1972. OECD Ministers adopt guiding principles for environment policy, Press/A(72)23, Paris.

PFAFF, M. and A. PFAFF. 1970. Grants Economics: An evaluation of government politics. Wayne State University, Detroit.

POUPARDIN, D. , P. M. RINGWALD et B. WOLFER, 1972. La contribution des économistes à l'étude de la disparition des espaces naturels périurbains, Ann. Econ. Social. Rur. 1.

SCOTT, A. , 1972, this book, p. 7.

SECKLER, D. W. , 1966. On the uses and abuses of economic science in evaluating public outdoor recreation, Land Econ. Nov.

SMETS, H. , 1972, this book, p. 75.

STEIN, R. E. , this book, p. 285.

GUIDELINES FOR THE GUIDELINES
ON THE QUESTION OF PHYSICAL INTERNATIONAL
EXTERNALITIES

by

S. Ch. Kolm

CEPREMAP, Paris, France

I

Relations between nations are an evolving phenomenon. We should
consider more what they will be than what they were. The trend seems
to be towards a slow decline of the nation-states, especially if we consider that the relations that interest us are those among Western European countries, on the one hand, and between the U.S. and Canada, on
the other. Economic and political integration, labour and leisure (tourism) mobility, and multinational corporations are on the increase.
Therefore, we will follow the trend of events if we press for person-
oriented solutions rather than nation-oriented ones. Economic inter-
dependency and integration will anyway force this option upon us. Along
the American Great Lakes and European rivers, many a firm which pol-
lutes a neighbouring country belongs to nationals of this victim country.

Consider also the non-discrimination principle: "do not do to a
foreigner what you would not do to a national", or "treat a foreigner as
you would have treated a national", or "treat agents irrespective of
their nationality". This principle is in many respects at the same time
good ethics, good economics, and good politics. But it is enough to note
that, applied to our field, it is just an extension to environmental questions of the basic EEC (or GATT) principle for trade and labour: to
adopt it, is, in this sense, a mere matter of consistency. Of course,
its application raises problems; but not insoluble ones, it seems.

If the aim is welfare, and since welfare is defined for persons and
not for "nations", person-oriented solutions will always be more directly to the point than nation-oriented ones. For instance, optimum solutions require that the right measures be associated with transfers. But
it is not government-to-government transfers which are relevant, but
transfers between persons of different nationalities. To suppose that
the government will forward the transfers to or from the right persons
is a bold assumption. In fact, inter-government transfers will not
generally be felt as acceptable, whereas inter-personal compensation
often will be, even between persons of different nationalities (a government can be a party to a transfer only if most of the nationals are concerned - as in the case of war reparations - but causes and effects of
transfrontier pollution are generally much more specific).

This shows another line of strategy: we should systematically
advocate the institution of international transfers (between persons of
different nationalities, or between persons and international - or a-
national but public - bodies).

II

It is when this advice is not acceptable that we would have recourse to national governments. And it is when they do not exchange money that we would have to advocate using other problems' parameters as means of exchange. At the limit, all pending international questions would be discussed together between the governments. And we may go further in allowing triangular governmental arrangements, and so on. The aim is to create enough variables to make up for the absence of direct money transfers.

But there are at least two drawbacks to this method: an increase in the costs of transaction, discussion, contact, etc., and an implicit assumption that the intra-national mechanisms correctly allocate the result to the right persons.

However, governments will spontaneously tend to use bargaining on these other parameters to settle the question under consideration. Therefore, the whole system of international relations may be affected, and an international body has to look into this process. Its aim would be to set orderly "rules of the game", in particular by defining the set of parameters (problems) available for bargaining about transfrontier pollution. A priori international agreements on these rules will permit a better further settlement of problems. How must we make this choice?

One possibility would be to forbid that any other question be settled jointly with the one under consideration, in order to force countries to resort to money compensations. But if they do not do this (or accept the rule), something else must be chosen. The principle would be that the set of chosen parameters should offer enough possibilities to be a complete substitute for payments in cash.

But we will certainly want to forbid some instruments. For instance, a recommendation not to use military threats to settle international pollution disputes would certainly be accepted! But we can go further. For instance, we would forbid any instrument the use of which runs contrary to other international policy aims of the group of nations under consideration. In particular, we would forbid the use of threats about international trade (tariffs, quotas, etc.). Note that anyway, within the EEC the use of such weapons would violate the EEC treaties. The rule will also be valid for capital and personal transnational movements.

Of course, it would be helpful if, in addition to parameters, bargaining is provided with starting situations or threats (to be implemented in the absence of agreement). Nature gives all the power to the polluter ("wildcat allocation") and, therefore, in the case of a unidirectional nuisance, the natural threat situation is completely unbalanced in favour of the polluting nation. Justice calls for a correction of nature in this respect (so it may be thought). A possibility would be to create a presumption that the reverse rule would have to be applied. For this, international organisations would promote a "polluter-pays principle" saying that the polluter is liable for all damage. This threat would not be complete and sure because international organisations have no means to enforce it upon nation-states. But it would not be without influence (moral pressure, existence of other questions in which the polluter has an interest, etc.). This would be a countervailing threat, compensating natural inequality. It is, of course, in the nature of a threat that there will be no attempt to implement it if the parties agree between themselves.

The present OECD "polluter-pays principle" (that is, pollution abatement, but not damage compensation, is charged to the polluting country) cannot play this role. Of course, the polluting country can often hardly avoid incurring abatement costs; for instance, when abatement is obtained by abstaining from polluting or from locating on the critical river it is difficult to compensate all would-be polluters. And in the many cases where damage is difficult to measure, a polluter's liability is not a precise instrument. But these two uncertainties (on abatement and damage) make the optimal (least global cost) pollution level also uncertain, and this principle makes the interests of these two nations on this variable definitely conflicting. Therefore, untruthful arguments will probably be used: this makes it very unlikely that international efficiency will be achieved. Consequently, the resulting situation will be such that there exist others in which everybody is better off.

Anyway, this OECD principle will often not be accepted in practice. It seems to imply, in addition to what it says, that the degree of pollution must be the one which minimizes global costs (abatement plus damage), i. e. that efficiency must prevail. But at this level it is the marginal costs of damage and abatement which must be equal, not the total ones. The latter can still be widely different. And in this case, agreement will generally not be reached at this efficient level, but at another, inefficient one.

In order to give the possibility of reaching efficient international agreement in all cases (and not only in fortuitous ones), another variable must be brought into the parameters of bargaining, in addition to pollution levels. This is the role of money payments and other variables discussed above. But their acceptability is limited by the fact that they do not seem to be closely linked to the environment question under discussion itself. A device which is equivalent to money payments but which is closely tied to the problem, would be to choose as a second parameter in the bargaining the share of global cost (abatement plus damage) which will be paid by each nation. Then, once this is chosen, the interests of the two parties will coincide in choosing the pollution level which makes this global cost minimum: efficiency will be guaranteed. It is easy to show that for all agreement on the pollution level with the OECD "polluter-pays principle" - except for the improbable unique efficient one under this rule - there exists a sharing of global damage associated with another pollution level which makes everybody better off. MM. Smets' and Muraro's studies are concerned with this point.

On the figure, the variables on the axes are the pollution level P and the payment t from the polluter to the victim (whether they are "nations" or private agents of different nationalities). When t is negative, -t is a payment from the victim to the polluter. $I_p(I_p^1, I_p^2, I_p^3)$ are indifference curves of the polluter. $I_v(I_v^1, I_v^2, I_v^3)$ are indifference curves of the victim. Points on the P axis correspond to no payment. A is the best of these points for the polluter, that is, "wildcat allocation". Line C is the locus of efficient situations; at each of its points, and I_p and I_c are tangent. The intersection B of line C with the P axis represents the situation of efficient pollution without payment. This is the OECD point; efficiency, polluter pays abatement, victim pays damage. But, in this situation, in the figure's case the cost to the polluter, BL, is much larger than the cost to the victim, BK. Therefore, the polluter will probably not agree to incur these abatement costs. Still without payment

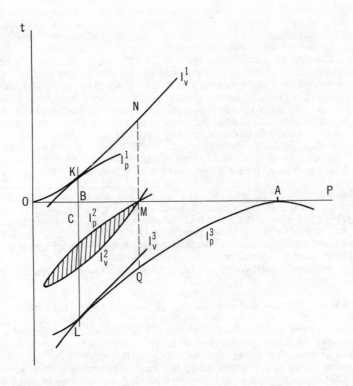

the countries may agree on the pollution level OM, where polluter's costs (MQ) and victim's costs (MN) are of the same order of magnitude. But, then, there exist many situations in which both are better off: all all those represented by points in the shaded area between indifference curves I_p^2 and I_v^2 passing through M. And some of these situations are efficient (points on C). However, all these situations require some money payment.

III

These bargainings and the implementation of the outcome are conveniently carried out by an <u>ad hoc</u> organisation, a "Pollution Control Authority" (PCA), specific to the problem and area. This PCA would be an international body, but even inside nations pollution problems (water in particular) usually go beyond or cut across traditional jurisdictions, so that PCA's have had to be created (River Basin Agencies). In France, PCA's have found wide favour and fast implementation (some unfairly say this is due to a national habit to tackle problems with institutions rather than with solutions).

The scope of a PCA is determined by the JJR (Jean-Jacques Rousseau) Principle: "he who is concerned by a decision must take part in it, and he who is not must not". In the present case, "concerned" means a direct creator or a direct victim of the physical externality. This principle is necessary to internalize this externality. The idea to favour person-oriented solutions over nation-oriented ones has a definite application here: the "legislative" body of the PCA must consist not of the national governments but of the citizens (or firms) concerned or, if this creates excessive policy-making costs, of the relevant local governments. The composition of this body could be similar to that of "Basin Committees" in France. Because of the present public finance circuits and taxation rights, national governments could also be represented, their place being that of the "representatives of the Administration" in French River Basin committees. However, a still better idea would be to exclude them in order to induce a change in these circuits and rights.

The principle would be that a PCA should internalize only a physical externality, and not market (efficient and purely distributional) ones, since the latter would be taken care of by "as pure as possible" redistribution transfers. How should it take into account related economic imperfections: inadequately solved other externality (environment) problems, market imperfections, etc.? A possibility would be to "play" second best. Another one would be to imitate Electricité de France's application of the marginal cost pricing principle. That is to ignore these imperfections in action, while claiming that they should be corrected. This latter behaviour is an application of Emmanuel Kant: act as if the principle of your action, if it were to become the general rule, would lead to the best of worlds.

Finally, adoption of the person-oriented view in international externalities would also be a means towards a higher aim which is to correct one of the major defects of the present world: the excessive concentration of collective decision-making at the nation-state level. Application of the JJR principle shows that there is almost no problem for which this level is the best.

TRANSNATIONAL POLLUTION
AND INTERNATIONAL LAW

by

Christopher Bo Bramsen

Legal Department, Danish Ministry of Foreign Affairs,
Lecturer in International Law, University of Copenhagen,

Denmark

I. INTRODUCTION

In recent years the protection of human environment has received increasing attention at the national as well as at the international level. Several conferences on pollution have been held, and international organisations have more and more been concerned with drafting regulations to prevent pollution from various sources. The problems of pollution was one of the main subjects discussed at the United Nations Conference on the Human Environment held in Stockholm in June of this year under the motto "Only One Earth"(1). IMCO is planning to convene a conference in 1973 at which marine pollution will be among the topics for deliberation(2), and the United Nations Committee on the Seabed is preparing a conference on the law of the sea at which the preservation of the sea (including measures to prevent pollution) will be on the agenda(3).

This paper discusses whether international law in general imposes upon States any obligation to prevent transnational pollution,[1] in other words to prohibit such activities within their national jurisdiction (including ships and aircraft) that might cause damage by pollution in areas outside their national jurisdiction.

II. POLLUTION AND INTERNATIONAL LAW

The amount of industrial refuse and wastes has grown at such a rate that disposal now presents a grave risk of pollutants transcending national frontiers and having deleterious effects in the territories of other States as well as in areas beyond national jurisdiction, such as the high seas and the air above. According to several global and regional treaties, States have assumed legal obligations to prevent transnational pollution. Only the States parties to these treaties are, of course, bound by such obligations. The all-important question is, however, whether there exists any general rule of international law under which it is incumbent upon States to stop transnational pollution(4).

Taking into consideration the growing interdependence of States, the development of industrial production, the mounting risks of pollution,

---

1. Transnational pollution covers more than transfrontier pollution, see "Draft Guiding Principles Concerning Transfrontier Pollution", paragraph 3.

the rise in consumption and the steadily increasing figure of world population it is possible that the application of the principles of good neighbourly relations(5) has developed into a prohibition in international law against transnational pollution.

The law of neighbourly relations, known from municipal law, was invoked in an international context as early as 1941 in the Trail Smelter case(6) submitted for arbitration by the United States and Canada. In this case, in which the United States claimed compensation from Canada for damage caused by emission of fumes from a Canadian factory, the tribunal held that "under the principles of international law, as well as of the law of the United States, no State has the right to use or permit the use of its territory in such manner as to cause injury by fumes in or to the territory of another or the properties or persons therein, where the case is of serious consequence and the injury is established by clear and convincing evidence"(7).

In its judgement on the Corfu Channel case the International Court of Justice referred to "every State's obligation not to allow knowingly its territory to be used for acts contrary to the rights of other States"(8).

Several international jurists also point to the aspect of neighbourly relations in the context of responsibility for water and air pollution, changes of water courses, nuclear test explosions, etc. and refer to a duty of non-interference established by customary international law, generally stated in the maxim: sic utere tuo ut alienum non laedas(9). (Use your own property so as not to injure your neighbour's).

Oppenheim-Lauterpacht(10) refer to this principle as "one of those general principles of law recognized by civilized States which the Permanent Court is bound to apply by virtue of Article 38 of its statute. However, the extent of the prohibition of abuse of rights is not at all certain. It is of recent origin in the literature and practice of International Law, and it must be left to international tribunals to apply and develop it by reference to individual situations".

In the preamble of the United Nations Declaration on Principles of International Law concerning Friendly Relations among States in Accordance with the Charter of the United Nations(11), the General Assembly recalls that "the people of the United Nations are determined to practice tolerance and live together in peace with one another as good neighbours".

In concordance with this view Max Sørensen(12) says: "Malgré ces incertitudes et difficultés il paraît justifié de conclure que l'exercice de la compétence territoriale de l'Etat est subordonné à un principe de bon voisinage international". (In spite of these uncertainties and difficulties, it seems reasonable to conclude that the exercise of a state's territorial sovereignty is subordinate to the principle of international good neighbourliness.)

Andrassy(13) states that "un principe général de droit international interdit à un Etat de faire sur son territoire des travaux, qui causeraient un préjudice grave au territoire d'un autre Etat". (It is a general principle of international law that no state may carry out on its own territory activities liable to cause grave damage to the territory of another state.) See also Thalmann(14) who points out that "Nachbarliches Völkerrecht verbietet den Staaten gewisse Handlungen im Grenzraum", and mentions as an example that "Verschmutzungen internationaler Gewässer, welche von schädlichen Folgen auf dem Gebiete des beeinträchtigten Staates begleitet sind, stellen nachbarrechtliche verbotene Handlungen dar". (According to international law the rules of neighbourly relations prohibit States from carrying out certain activities in their border-areas. Pollution of international waters having harmful effects in the territory of the concerned State is forbidden by the law of neighbourly relations.)

In a draft declaration on rights and obligations of States(15), the International Law Commission agreed that "every state has the duty to ensure that conditions prevailing in its territory do not menace international peace and order".

Expressing the same opinion at the UN Conference on Water Pollution in Europe which was held in Geneva in 1961, E.J. Manner observed (16) that "an extension of the consequences of pollution into another State's territory must be regarded as violation of that State's integrity or as a certain kind of interference in international law". The Conference pointed out that "as far as customary international law is concerned its content is as yet not clearly defined and universally agreed upon and might vary in time as well as from one group of States to another"(17).

In Survey of International Law, 1971(18), this subject is mentioned: "There is of course a certain amount of customary law which may be referred to in this context and a number of cases relating to the application of the general principles of international law which may be invoked whilst treaties contain provisions which may be pertinent. (...) Nevertheless it is understood that the task confronting the international community entails a development of essentially new law, on what may eventually prove to be a considerable scale and not merely the codification of existing legal rules and practices. "

The United Nations Conference in Stockholm agreed on a Declaration on the Human Environment(19), from which the following paragraph is quoted:

"21.   States have, in accordance with the Charter of the United Nations and the principles of international law, the sovereign right to exploit their own resources pursuant to their own environmental policies, and the responsibility to ensure that activities within their jurisdiction or control do not cause damage to the environment of other States or of areas beyond the limits of national jurisdiction. "(20)

Similar views are expressed in a resolution of the United Nations General Assembly(21) which states that "each country has the right to formulate, in accordance with its particular situation and in full enjoyment of its national sovereignty, its own national policies on the human environment, including criteria for the evaluation of projects. In the exercise of such right and in the implementation of such policies due account must be taken of the need to avoid producing harmful effects on other countries. "

Thus it seems to be a general assumption that the freedom of action of States is subject to restrictions following from the law of neighbourly relations, and that such restrictions are of particular relevance in matters of pollution. A comparison of international jurisprudence and general principles of law, on the one hand, and the legal concepts reflected in preambles of international conventions and in recommendations, resolutions and declarations, on the other, would indicate the extent to which a prohibition on transnational pollution has been established under international law for each category of pollution.

III.   CATEGORIES OF POLLUTION

A review of the rules of international law relating to pollution can be based on a classification according to the nature of the pollutant, the medium in which pollutants are spread, or the nature of damage caused by pollution. This paper primarily considers the media by which pollutants are transported across frontiers, namely sea, water and air, except in the case of radioactive and noise pollution where classification according to pollutant has proved to be the most expedient approach for purposes of legal regulation.

A.   MARINE POLLUTION (22)

1.   General Principles

The legal provisions governing the acts of States on the high seas are based on the principle of freedom of the seas. Article 2 of the Geneva Convention of 1958 on the High Seas(23) defines the freedom of States and lists the following restriction: "These freedoms, and others which are recognized by the general principles of international law, shall be exercised by all States with reasonable regard to the interests of other States in their exercise of the freedom of the high seas".

The Geneva Convention on the Continental Shelf(24) in Article 5, paragraph 1 also points to this obligation of any State: "The exploration of its natural resources must not result in any unjustifiable interference with navigation, fishing or the conservation of the living resources of the sea, nor result in any interference with fundamental oceanographic or other scientific research carried out with the intention of open publication. "

The General Assembly of the United Nations has repeatedly expressed concern at the increasing pollution of the marine environment, as is shown in the following excerpt from the preamble of a resolution adopted in 1968(25):

"Mindful of the threat to the marine environment presented by pollution and other hazardous and harmful effects which might result from exploration and exploitation of the areas under consideration ... ".

In a resolution adopted in 1969(26) the General Assembly requested the Secretary General to obtain the views of Member States on the desirability and feasibility of an international treaty or treaties relating to the prevention and control of marine pollution. Although the governments who replied to the question put by the Secretary General did not entirely agree on the technical and legal methods and procedures concerning the prevention of pollution, the report of the Secretary General(27) states that "the general opinion expressed in the replies received is that marine pollution presents an increasing danger to the marine environment and that steps should therefore be taken at the intergovernmental level towards its prevention and control." These replies seem to indicate that States consider themselves bound to respect a ban on harmful marine pollution.

In its Declaration of Principles Governing the Sea-Bed and the Ocean Floor, and the Subsoil Thereof, beyond the Limits of National Jurisdiction(28), the General Assembly declares that "with respect to activities in the area and acting in conformity with the international régime to be established, States shall take appropriate measures for and shall co-operate in the adoption and implementation of international rules, standards and procedures for, inter alia:

a) The prevention of pollution and contamination, and other hazards to the marine environment, including the coastline, and of interference with the ecological balance of the marine environment;

b) The protection and conservation of the natural resources of the area and the prevention of damage to the flora and fauna of the marine environment."

In November 1971 in Ottawa the United Nations Intergovernmental Working Group on Marine Pollution adopted certain general Guidelines and Principles for the Preservation of the Marine Environment(29). Principle No. 1 seems to be of interest in this context: "Every State has a duty to protect and preserve the marine environment and, in particular to prevent pollution that may affect areas where an internationally shared resource is located." Principle No. 5 provides that "States should assume joint responsibility for the preservation of the marine environment beyond the limits of national jurisdiction."

The Working Group adopted the following definition: "Marine pollution is defined as the introduction by man, directly or indirectly, of substances or energy into marine environment (including estuaries), resulting in such deleterious effects as harm to living resources, hazards to human health, hindrance to marine activities, including fishing, impairment of quality of use of seawater, and reduction of amenities."

The Stockholm Conference agreed on the following principle(30): "7. States shall take all possible steps to prevent pollution of the seas by substances that are liable to create hazards to human health, to harm living resources and marine life, to damage amenities or to interfere with other legitimate uses of the sea".

2. The Rights and Duties of Coastal States

The sea beyond national jurisdiction must be regarded as res communis omnium. It follows from the rights and duties established for coastal states in recent years in regards to transnational pollution, that various activities on the high seas and in the territories of adjacent coastal States are subject to restrictions defined in terms of consideration for the interests of other States.

As for the rights of coastal States, the Convention of 1958 on the Territorial Sea and the Contiguous Zone(31) contains this provision in Article 24, paragraph 1: "In a zone of the high seas contiguous to its territorial sea, the coastal State may exercise the control necessary to prevent infringement of its customs, fiscal, immigration or sanitary regulations within its territory or territorial sea;"

In line with this provision, the Geneva Convention on Fishing and Conservation of the Living Resources of the High Seas(32) stipulates, in article 6, paragraph 1, that "a coastal State has a special interest in the maintenance of the productivity of the living resources in any area of the high seas adjacent to its territorial sea.", and article 7 of the same Convention continues: "Having regard to the provisions of paragraph 1 of article 6, any coastal State may, with a view to the maintenance of the productivity of the living resources of the sea, adopt unilateral measures of conservation appropriate to any stock of fish or other marine resources in any area of the high seas adjacent to its territorial sea, provided that negotiations to that effect with the other States concerned have not led to an agreement within six months."

The rights of coastal States are also emphasized in the International Convention relating to Intervention on the High Seas in cases of Oil Pollution Casualties mentioned below.

The above mentioned principles relating to the seabed(33) emphasize the rights of coastal States in these terms:

"12. In their activities in the area, including those relating to its resources, States shall pay due regard to the rights and legitimate interests of coastal States in the region of such activities, as well as of all other States, which may be affected by such activities. Consultations shall be maintained with the coastal States concerned with respect to activities relating to the exploration of the area and the exploitation of its resources with a view to avoiding infringement of such rights and interests.

"13. Nothing herein shall affect: (a) The legal status of the waters superjacent to the area or that of the air space above those waters; (b) The rights of coastal States with respect to measures to prevent, mitigate or eliminate grave and imminent danger to their coastline or related interests from pollution or threat thereof or from other hazardous occurrences resulting from or caused by any activities in the area, subject to the international régime to be established."

Reference to the special interests of coastal States is made also in Principle 21 of the Ottawa-document(34) which reads: "Following an accident on the high seas which may be expected to result in major deleterious consequences from pollution or threat of pollution in the sea, a coastal State facing grave and imminent danger to its coastline and related interests may take appropriate measures as may be necessary to prevent, mitigate, or eliminate such danger, in accordance with internationally agreed rules and standards."

While it would seem reasonable to grant considerable rights to coastal States to safeguard their interests, unreasonable restrictions on the freedom of States on the high seas would be adverse to the interests of general shipping. Some States are unilaterally contemplating the establishment of pollution zones contiguous to their territorial waters(35). The weighing of interests which is required for finding a solution to these problems could most appropriately, it seems be performed by reference to an international forum. It is unquestionable, however, that ecological factors show the growing need for tightening the obligations of States in order to prevent damage by pollution in the territories of coastal States.

It seems also established that it is the duty of coastal States not to pollute common areas such as the high seas.

The Geneva Convention on the Continental Shelf(36) imposes, in article 5, paragraph 7, the following obligation upon States operating installations on the continental shelf: "The coastal State is obliged to undertake, in the safety zones, all appropriate measures for the protection of the living resources of the sea from harmful agents."

Principle No. 17 of the Ottawa-principles(37) clearly underlines the obligation of coastal States to prevent transnational pollution. It reads: "In addition to its responsibility for environmental protection within the limits of its territorial sea, a coastal State also has responsibility to protect adjacent areas of the environment from damage that may result from activities within its territory".

The preparatory document on pollution to the United Nations Conference in Stockholm(38) points out that "the oceans are of use to all mankind, and pollutants entering the oceans in one area may have undesirable effects far from where they enter; coastal nations have a particular interest in and responsibility for these matters, but because most marine pollution arises from polluting activities on land, often far from the sea, all States should be involved in a common effort to manage the marine environment so as to preserve its quality and conserve its resources."

3. Oil Pollution

At an early stage the discharge into the sea of oil from ships became subject to national as well as international regulation. In 1954, the International Convention for the Prevention of Pollution of the Sea by Oil(39) was signed in London under the auspices of IMCO. Amendments in 1962(40) and 1969(41) made the ban on discharge of oil into the sea almost complete, and it is now no longer confined to certain geographical zones.

The Geneva Convention of 1958 on the High Seas(42) clearly underlines, in Article 24, the obligation of States to pay regard to the environment and to the interests of other States: "Every State shall draw up regulations to prevent pollution of the seas by the discharge of oil from ships or pipelines or resulting from the exploitation and exploration of the seabed and its subsoil, taking account of existing treaty provisions on the subject." This provision is generally regarded as codifying existing customary law.

The Torrey Canyon accident of 1967 provided an incentive for several States to enter into regional agreements on assistance in cases of oil pollution of the sea(43). These agreements clearly demonstrate the application of the principle of respect for the interests of other States.

A conference held in Brussels in 1969 under the auspices of IMCO resulted in two conventions on oil pollution. The International Convention relating to Intervention on the High Seas in Cases of Oil Pollution Casualties(44) emphasizes the relationship between the rights and duties deriving from the principle of freedom of the high seas:

"The States Parties to the present Convention, conscious of the need to protect the interests of their peoples against the grave consequences of a maritime casualty resulting in danger of oil pollution of sea and coastlines,
convinced that under these circumstances measures of an exceptional character to protect such interests might be necessary on the high seas and that these measures do not affect the principle of freedom of the high seas, have agreed as follows:

"Article I.

1.   Parties to the present Convention may take such measures on
the high seas as may be necessary to prevent, mitigate or eliminate
grave and imminent danger to their coastlines or related interests
from pollution or threat of pollution of the sea by oil, following upon
a maritime casualty or acts related to such a casualty, which may
reasonably be expected to result in major harmful consequences. "

Another IMCO-Convention of 1969, relating to Civil Liability for
Oil Pollution Damage(45) was followed in 1971 by a Convention on the
Establishment of an International Compensation Fund for Oil Pollution
Damage(46).

It is clear from the foregoing that oil pollution has been the subject
of much attention. There seems to be broad agreement in shipping circles
as well as among environmentalists that every possible effort should be
made to fight this form of pollution. But while there is agreement on the
need for prohibitions of various kinds, it has not yet proved feasible to
implement effective sanctions. On this point J. P. Queneudec(47) says:
"Il ne suffit pas aujourd'hui d'élaborer un minimum de règles communes
dans le but d'assurer la sécurité de la navigation, la sauvegarde de la vie
humaine en mer, et la conservation des ressources marines. Il est de-
venu indispensable d'instituer une véritable police internationale de la
mer, afin de préserver ce patrimoine commun de l'humanité contre les
dangers auxquels l'exposent les progrès de la science et de la technique.
Cela ne peut se faire qu'au prix d'une rénovation du droit de la mer, par
l'introduction de sérieuses limitations au principe de la liberté de navigation et à la loi du pavillon, qui se trouveraient ainsi "désacralisés". "
(These days it is no longer sufficient to formulate a few common rules
in order to ensure the safety of navigation, the security of persons at
sea and the conservation of marine resources. It has become essential
to organise an actual international police force for the purpose of protecting this common heritage of mankind against the dangers to which it
is exposed through scientific and technical evolution. Only a complete
revision of the law of the sea can achieve this end by severely restricting
and curtailing the sacred principles of freedom of navigation and of the
law of the flag.)

4.   Ocean Dumping

Pollution caused by ocean dumping of industrial wastes was one of
the main items on the agenda of the Conference on Marine Pollution and
the Effects thereof on the Living Resources of the Sea and on Fishing,
held in Rome in December 1970 under the auspices of the FAO. The
conference was attended by a large number of experts from all member
States of the United Nations and its recommendations have served as a
guideline in subsequent international work on marine pollution.

In the discussion of the rights of States regarding dumping on the
high seas, a conflict arises between the principle of freedom of the sea
and the principle of banning transnational pollution. On this point
Queneudec(48) says: "Chaque Etat a certes le droit, au nom de la liberté
de la mer, de rejeter dans les océans certains résidus; mais il a égale-
ment l'obligation de respecter le principe de l'interdiction de pollution
des eaux maritimes. Nul doute qu'il peut y avoir ici un conflit entre ces
deux principes internationalement reconnus. " (Because of freedom of
the seas every State of course has the right to discharge some waste into

the ocean; but it likewise has a duty to respect the principle prohibiting pollution of maritime waters. Undoubtedly a conflict may arise from the confrontation of these two internationally recognized principles.)

In February 1972 twelve Western European States agreed on a Convention for the Prevention of Marine Pollution by Dumping from Ships and Aircraft(49). The Convention, which covers the North-East Atlantic area, deals primarily with ocean dumping, but it is worth noting in the preamble the reference to neighbourly relations: "Considering that the States bordering the North-East Atlantic have a particular responsibility to protect the waters of this region ...". In article 1 the Contracting Parties pledge themselves "to take all possible steps to prevent the pollution of the sea by substances that are liable to create hazards to human health, to harm living resources and marine life, to damage amenities or to interfere with other legitimate uses of the sea."

In the draft articles of a Convention for the Prevention of Marine Pollution by Dumping produced by the Intergovernmental Meeting on Ocean Dumping in Reykjavik in April 1972(50) a similar pledge is made in article 1. Of relevance to the present case, however, is the fact that the preamble to these draft articles expressly mentions the obligation of States to pay regard to the interests of other States: "Recognizing that States have the responsibility to ensure that activities within their jurisdiction or control do not cause damage to the environment of other States or of areas beyond the limit of national jurisdiction ..."(51). Furthermore article X expressly mentions the prohibition against transnational pollution caused by dumping: "The Parties recognize that in accordance with the principles of international law States bear responsibility for damage to the environment of other States or to areas beyond the limits of national jurisdiction caused by dumping and undertake to develop procedures for the assessment of liability and for the settlement of disputes".

B. FRESH WATER POLLUTION(52)

Transmission of pollutants through international rivers is not a new problem. In numerous cases States have taken measures under regional agreements to settle any disputes that may arise(53). There exists no universal convention in this field, but various organisations have tried to define water pollution and to formulate certain principles with respect to the use of international rivers.

As early as in 1911 the Madrid Declaration of the Institute of International Law(54) laid down the following rule: "Toute altération nuisible de l'eau, tout déversement de matières nuisibles (provenant de fabriques, etc.), est interdit." (All alterations injurious to the water, the emptying therein of injurious matter (from factories, etc.) is forbidden).

At the Conference of the International Law Association held in Helsinki in 1966, experts on international law from approximately forty countries agreed upon the so-called Helsinki Rules on the Uses of the Waters of International Rivers(55). These rules, which represent an attempt to codify the general principles of international law in this respect, contain important provisions about water pollution. Water pollution is defined in article IX as "any detrimental change resulting from human conduct in the natural composition, content, or quality of the waters of an international drainage basin". According to the rules laid down in article X, "a State must prevent any form of water pollution or any increase in the degree of existing water pollution in an international drainage basin which would cause substantial injury in the territory of a co-basin State, and should take all reasonable measures to abate existing water pollution in an international drainage basin to such an extent that no substantial damage is caused in the territory of a co-basin State". These rules apply to "water pollution originating within a territory of the State, or outside the territory of the State, if it is caused by the State's conduct".

Similar criteria are applied with respect to claims for damages arising between neighbouring States, where the beneficial effects of an act are weighed against its harmful effects. The question of proportionality, which is a leading theme in the rules of law governing relations between neighbouring States, is discussed by O'Connell(56): "Is the damage done to neighbouring States as a result of the discharge of industrial waste into rivers which traverse several territories disproportionate to the total benefit to those territories of the industrial activity in question? The most satisfactory way of solving the problem of river pollution is by joint action on the part of riparian States, but in the absence of agreement it can be argued that there is an obligation on States to take measures against excessive discharge of waste into watercourses which supply international rivers, and particularly against the discharge of extremely hazardous substances."

European collaboration in the combatting of water pollution has been gradually increasing. The ECE definition of water pollution(57) states that "a river is polluted when the water in it is altered in composition or condition, directly or indirectly as the result of the activities of man, so that it is less suitable for any or all of the purposes for which it would be suitable in its natural state".

The European Water Charter(58) points specifically to the interests and welfare of man: "To pollute water is to harm man and other living creatures which are dependent on water". It also emphasizes that "water knows no frontier, as a common resource it demands international co-operation."

These principles defined in the Water Charter have been further developed in the Council of Europe Draft Convention on Protection of Fresh Waters against Pollution(59) by the incorporation - in a chapter dealing with damage caused by pollution - of provisions imposing responsibility on States for damage caused on the territory of other States. Any Contracting Party which commits, authorizes or tolerates acts that, by polluting water, cause damage to others, is according to draft article 13 responsible towards the other Contracting Party or Parties on whose territory the damage is sustained. The theory of neighbourship rights and duties is reflected in draft article 14: "In case of a sudden increase in the level of pollution, even if unforeseen and not authorized or tolerated, the Contracting Parties interested in a same drainage basin are under a duty to take immediately, unilaterally and, if necessary, jointly all measures in their power to prevent consequences causing damage or to limit its extent, and, if they fail to do so, they will be liable to the other Contracting Party or Parties on whose territory the damage is sustained."

The aspect of pollution is also mentioned in the preamble of the 1968 European Agreement on the Restriction of the Use of certain Detergent in Washing and Cleaning Products(60): "Whereas it is becoming increasingly necessary to secure harmonization of the Laws on the control of fresh water pollution ... ".

In the Lac Lanoux case(61) reference was made to the principles governing relations between neighbouring States. The award stated that: "On aurait pu soutenir que les travaux auraient pour conséquence une pollution définitive des eaux du Carol, ou que les eaux restituées auraient une composition chimique ou une température, ou telle autre caractéristique pouvant porter préjudice aux intérêts espagnols", (it could have been argued that the works could bring about a definite pollution of the waters of the Carol or that the returned waters could have a chemical composition or a temperature or some characteristic which could injure Spanish interests). (62)

On the basis of this award Brierly-Waldock(63) formulate various general rules, among them one to the effect that "a state in principle is precluded from making any change in the river system which could cause substantial damage to another state's right of enjoyment without that other state's consent. "

C. AIR POLLUTION

In the Trail Smelter decision(64) the question of pollution was dealt with in substance at the international level, seemingly for the first time.

A privately owned Canadian company emitted through its daily smelting of zinc and lead ores, enormous concentrations of sulphur dioxide fumes causing substantial damages in the United States. As mentioned above(65) the tribunal concluded that "no State has the right to use or permit the use of its territory in such a manner as to cause serious injury by fumes in or to the territory of another". The tribunal also held that Canada was responsible under international law for the conduct of the Trail Smelter. "It is therefore the duty of the Government of the Dominion of Canada to see to it that this conduct should be in conformity with the obligation of the Dominion under international law as herein determined. "(66)

As pointed out in the document on pollution(67) prepared for the UN Conference on the Human Environment, pollutants may be transported over great distances through the atmosphere and affect areas hundreds of kilometres away: "Many cases of such transport have been documented, for example: dust picked up by wind storms over North Africa falling over Europe, smoke from forest fires in North America being visible in Europe and products of industrial processes on the east of North America being detected at a distance of 300 kilometres over the Atlantic Ocean. "

In the Swedish case study for the Environment Conference "Air Pollution across National Boundaries"(68) it is emphasized that "the probable average time of a few days during which sulphur pollutants remain airborne means that the deposition in any one place will be dependent on the emissions within a surrounding area with a radius of one to two thousand kilometres. The problem is thus an international one. " The consequence of this is "that it will be necessary in many cases to have international (or in certain cases bilateral) agreements drawn up to cope with it."

International organisations have established more or less detailed guidelines and regulations to prevent air pollution across national frontiers.

In 1958 the United Nations Conference on the Law of the Sea established in the Convention on the High Seas(69) that "All States shall co-operate with the competent international organisations in taking measures for the prevention of pollution of seas or air space above, resulting from any activities with radioactive materials or other harmful agents."

The NATO Committee for Challenges to Modern Society has adopted an air pollution resolution(70) to the effect that all member States of NATO shall "take cognizance of the need for dynamic efforts toward enhancing and maintaining the quality of the air; endeavour to use, where appropriate, for setting up national air quality management programmes, the systems methodology generated by the CCMS pilot study; and promote co-operation among members when air pollution problems are common to national boundaries."

OECD has initiated a programme to measure the long-range transport of air pollutants(71). The Council of Europe is also dealing with problems of air pollution. Its declaration of 8th March 1968 contains the following definition: "Air is deemed to be polluted when the presence of a foreign substance or a variation in proportion of its components is liable to have a harmful effect or to cause nuisance". (71a)

The Trail Smelter decision seems to indicate quite clearly that harmful transnational air pollution is prohibited under existing rules of international law. Several international jurists hold the same view, among them Oppenheim-Lauterpacht(72) who declare:" A State is bound to prevent such use of its territory as, having regard to the circumstances, is unduly injurious to the inhabitants of the neighbouring State, e. g. as the result of working factories emitting deleterious fumes."

D.   RADIOACTIVE POLLUTION

Nuclear science has given man a new source of energy but also a source of harmful pollution by radioactive substances which may be transported through air, earth and water.

As mentioned above, the Convention on the High Seas(73) states that "All States shall co-operate with the competent international organisations in taking measures for the prevention of pollution of the seas or air space above resulting from any activities with radioactive materials or other harmful agents."

In an attempt to control radioactive pollution caused by the use of stationary or mobile nuclear plants, measures in the form of safeguards and rules on liability have been taken at the national as well as at the international level. International co-operation in this field under the auspices of IAEA, ENEA and IMCO has resulted in several conventions(74).

A problem of major international concern has been the elimination of radioactive waste. The Convention on the High Seas provides in article 25, paragraph 1, that "Every State shall take measures to prevent pollution of the seas from the dumping of radioactive waste, taking into account any standards and regulations which may be formulated by the competent international organisations".

In this connection the EURATOM Agreement(75) could also be quoted: "Each Member State shall provide the Commission with such general data relating to any plan for the disposal of radioactive waste in whatever form as will make it possible to determine whether the implementation of such plan is liable to result in the radioactive contamination of the water, soil or airspace of another Member State."

The States parties to the Oslo Convention on Dumping of 1972(76) could not agree to include radioactive waste in the lists of prohibited substances. Instead, the following provision was inserted in article 14: "The Contracting Parties pledge themselves to promote, within the competent specialized agencies and other international bodies, measures concerning the protection of the marine environment against pollution caused by oil and oily wastes, other noxious or hazardous cargoes, and radioactive materials."

A similar provision is found in article XI of the draft articles of the global dumping convention mentioned above under "Ocean Dumping". It is, however, still uncertain whether it will be possible in the final draft to include "High-level radioactive wastes" in the list of substances of which dumping is prohibited(77).

The numerous series of test explosions of nuclear weapons that have taken place in the post-war period have increased the risk of pollution inherent in released radioactivity. The Test Ban Treaty of 1963(78) mentions in the preamble "the desire (of the Contracting Parties) to put an end to the contamination of Man's environment by radioactive substances." And in article 1 of the Treaty the Parties undertake " to prohibit, to prevent, and not to carry out any nuclear weapon test explosion, or any other nuclear explosion, at any place under its jurisdiction or control: (a) in the atmosphere; beyond its limits, including outer space; or under water, including territorial waters or high seas; or (b) in any other environment if such explosion causes radioactive debris to be present outside the territorial limits of the State under whose jurisdiction or control such explosion is conducted."

In line with these provisions, the Stockholm Conference on the Human Environment(79) agreed on the following principle: "Man and his environment must be spared the serious effects of further testing or use in hostilities of nuclear weapons, particularly those of mass destruction."

The Test Ban Treaty has been followed up by a series of treaties(80) designed to prevent the geographical spread of nuclear weapons.

Nuclear explosions for peaceful purposes have not yet been covered under international treaties. Such explosions constitute a source of transnational pollution for which provisions should be laid down. There seems to be no doubt, however, that in this respect also, States are bound to have regard to other States, cf. Verdross(81): "Es besteht daher kein Zweifel darüber, dass Atomexplosionen, die auf fremde Staatsgebiete hinüberwirken, völkerrechtswidrig sind." (It is therefore unquestionable that nuclear explosions affecting the territories of foreign states are unlawful according to international law.)

E. NOISE POLLUTION (82)

Noise has become a serious problem in man's environment. Noise from traffic and factories has increased in the last years, but has not reached the point where it is of international concern.

For modern air transport, however, it has become necessary to work out standards for tolerable noise levels. The ICAO Convention on International Civil Aviation of 1949(83) regulates public international air law and postulates the fundamental principle that States have complete sovereignty over the air space above their territories. An Annex(84) to the Convention has recently been adopted by ICAO. It contains international standards and recommended practices related to aircraft noise certification, noise measurement and noise abatement operating procedures. This Annex deals only with certain types of subsonic aircraft.

The real danger of transnational pollution in this connection lies in the operation of supersonic aircraft that are able to cause several types of damages to property and hazards to health(85). It is feared that the exhausts from supersonic aircrafts discharged in the stratosphere might have serious effects on the climate. Some scientists also believe that supersonic aircrafts will cause a reduction in the atmospheric layer of ozone, allowing more ultraviolet radiation to reach the earth. Furthermore, when these aircraft reach a speed that is faster than the velocity of sound, a wide carpet of sonic booms sweeps across the surface of the earth underneath in a path of 40-60 kilometres on either side of the flight route. These booms and the noise produced in the take-off and landing phases of supersonic aircrafts have caused sleep disturbance and damages to property in numerous cases.

So far no international convention dealing with noise pollution produced by supersonic aircraft has been made, but several countries are preparing regulations prohibiting supersonic flight over their territories(86), and ICAO is working on international recommendations in this field.

The problem arises, however, whether a State is allowed under international law to prohibit a supersonic aircraft from flying outside its territory along the borders if the sonic booms produced by supersonic aircrafts cause injurious effects inside the territory of the State. As mentioned above under Section III A 2, a State has the right to establish a contiguous zone to its territorial sea to prevent infringement on certain regulations within its territory. It is not inconceivable that the principles reflected in the right to establish these zones and the theory of neighbourship rights could be invoked in the question of the legality of flights by supersonic aircrafts. The problem of possible restrictions in this field, which would also imply limitations on the freedom of the high seas(87), has not yet been solved by the international community(88).

272

IV. CONCLUDING REMARKS

The present survey of the sources of public international law including the legal attitudes put forward in international declarations, recommendations and resolutions demonstrates that rules of neighbourly relations do exist between States. The review seems to establish also that these rules are applicable in the field of pollution.

The conclusion is therefore that harmful transnational pollution is unlawful under international law(89). This rule can be based on various sources of law(90).

A large number of multilateral and bilateral treaties dealing with the law of pollution in various fields are now in force between States. In several cases the purpose of these treaties has been to codify existing international customary law. The International Law Association(91) and the Institute of International Law(92) have also based their work concerning pollution on international custom. That a prohibition against harmful pollution across boundaries is established under customary law can be deduced from the general practice of a great number of States, who in their municipal legislations within the different fields of pollution have prohibited emissions of harmful substances. The subjective element necessary to establish a customary rule, (the opinio juris sive necessitatis,) has manifested itself in a number of resolutions and declarations(93).

The classification of the sources of international law laid down in article 38 of the Statute of the International Court of Justice also includes "the general principle of law recognized by civilized nations". As mentioned above(94), the Roman principle of sic utere tuo ut alienum non laedas is often referred to in literature on pollution, and has found concrete expression in numerous river agreements which in turn have contributed to the formation of customary international law. (The concept of servitude and the doctrine of abuse of rights have also been invoked as principles covering the law of pollution but do not appear to have gained very wide acceptance (95).) The doctrine of neighbourship is supported by many international jurists(96) especially as it reflects both the principle of sic utere tuo and the interdependence of States.

Within international judicial decisions the Trail Smelter case has become the "leading case" in the field of pollution, and the principles laid down in the arbitral award seem to be accepted by almost all writers on the subject (97).

The general ban on transnational discharges of harmful pollutants underlines the imperative need to define what is to be understood by the term "harmful" in the different cases. It is therefore necessary that States in co-operation with the international organisations(98) work out "primary protection standards", "derived working limits" (99) and levels of acceptable pollution in order to clarify the extension of the above-mentioned ban. The UN Stockholm Conference recommended (100) "that in establishing standards for pollutants of international significance, governments take into account the relevant standards proposed by competent international organisations, and concert with other concerned governments and the competent international organisation in planning and carrying out control programmes for pollutants distributed beyond the national jurisdiction from which they are released. "

The obligations for States not only to refrain from causing harm to other States but also to promote active co-operation across the borders are also emphasized in the Ottawa-principles(101): "7. States should discharge, in accordance with the principles of international law, their obligations towards other States where damage arises from pollution caused by their own activities or by organisations or individuals under their jurisdiction and should co-operate in developing procedures for dealing with such damage and the settlement of disputes."

In the UN Resolution on Development and Environment adopted on 20th December 1971, the General Assembly "urges the international committees and the organisations of the United Nations System to strengthen international co-operation in the fields of environment, rational utilization of natural resources and preservation of adequate ecological balance(102)."

In another resolution of the same date, the General Assembly "recognizes the importance of ensuring that the global efforts in the field of the human environment be supplemented and made more effective by agreements at the regional or subregional levels(103)."

That the development of the law of pollution demands international co-operation is also pointed out in Survey of International Law(104): "States are now being impelled towards the adoption of a more active and deliberate approach to the development of international law than was formerly the case. Whereas international law was traditionally created largely by a series of bilateral acts, usually continued over an appreciable period, attempts are now made to tackle international problems on a more collective and regular basis." In this connection the following principles in the UN Declaration on the Human Environment(105) should also be quoted:

"24.   International matters concerning the protection and improvement of the environment should be handled in a co-operative spirit by all countries, big or small, on an equal footing. Co-operation through multilateral or bilateral arrangements or other appropriate means is essential to prevent, eliminate or reduce and effectively control adverse environmental effects resulting from activities conducted in all spheres, in such a way that due account is taken of the sovereignty and interests of all States.

"25.   States shall ensure that international organisations play a co-ordinated, efficient and dynamic role for the protection and improvement of the environment."

Protection of the environment can be secured to a certain extent through national legislation prohibiting completely or partially the discharge of a number of harmful substances. In these cases State responsibility for damage caused by harmful transnational pollution is assumed under the general principles of responsibility for internationally wronful(106) acts and omissions.

Certain activities, such as the operation of nuclear power stations and ships or the exploration and exploitation of the seabed and its subsoil entail serious risk of pollution resulting from accidents. The total benefit accruing from such activities may sometimes outweigh the risks involved, in which case these acts are in themselves considered lawful. The doctrine of neighbourship requires, however, that the international community reaches agreements on safety and control measures(107), sets up contingency plans and develops procedures for the settlement of disputes(108), and the assessment of liability(109). The question will have to be considered whether responsibility for damages caused by such dangerous and ultra-hazardous activities must be strict or absolute as opposed to the principle of negligence or fault applied in other activities involving the risk of harm(110). The problems concerning compulsory insurance, the establishment of a possible ceiling of pecuniary liability and the determination of exculpatory circumstances should also be dealt with in this connection. A general discussion on the various aspects of the question of State responsibility for accidents resulting from lawful activities seems, however, to go beyond the scope of the present paper.

FOOTNOTES

1. See UN Doc. A/RES 2398 (XXIII) of 3 December 1968, 2581 (XXIV)
    of 15 December 1969, 2657 (XXV) of 7 December 1970 and 2850
    (XXVI) of 20 December 1971.

2. See Resolution A.176 (VI) on marine pollution adopted by the As-
    sembly of the Intergovernmental Maritime Consultative Organisa-
    tion on 21 October 1969, calling for an international conference in
    1973 for the purpose of preparing a suitable international agree-
    ment for placing restraints on the sea, land and air by ships and
    other vessels or equipment operating in the marine environment.
    See also IMCO Doc. A 237 (VII) of 2 November 1971, where the
    IMCO Assembly decided "that the International Conference on
    Marine Pollution 1973 shall have as its main objectives the achieve-
    ment, by 1975 if possible but certainly by the end of the decade,
    of the elimination of the willful and intentional pollution of the seas
    by oil and noxious substances other than oil, and the minimization
    of accidental spills."

3. See UN Doc. A/RES 2340 (XXII) of 18 December 1967, 2467 (XXIII)
    of 21 December 1968, 2574 (XXIV) of 15 December 1969, 2749 and
    2750 (XXV) of 17 December 1970.

3a. Transnational pollution covers transfrontier pollution as well as
    pollution of the sea and pollution of countries by activities carried
    out in the sea (see also paragraphs 3 and 4 of "Draft Guiding
    Principles Concerning Transfrontier Pollution", p. 299 this book.

4. For a basic discussion on the law of the environment see Law,
    Institutions and the Global Environment, edited by J.C. Hargrove,
    Leiden, 1972.

5. For a general treatment on this subject, see J. Andrassy, "Les
    relations internationales de voisinage", Hague Recueil, vol. 79,
    1951 II, and Hans Thalmann, Grundprinzipen des modernen
    zwischenstaatlichen Nachbarrechts, Zürich 1951, see also C. W.
    Jenks, The Common Law of Mankind, London 1958, pp. 158-163,
    and A. Verdross, Völkerrecht, Wien 1964, p. 294.

6. Reports of International Arbitral Awards, vol. III, p. 1905.

7. Ibid., p. 1965. (All the underlining in this paper is the author's
    own.)

8. I.C.J. Reports, 1949, p. 22.

9. See e.g. the authors mentioned above in footnote 5.

10. Oppenheim, International Law, 8th ed. (edited by H. Lauterpacht)
    1955, p. 346.

11. UN Doc. A/RES 2625 (XXV) of 24 October 1970. See also the UN Charter, article 74 that refers to "the general principle of good-neighbourliness".

12. Max Sørensen, "Principes de droit international public", Hague Recueil, 1960 III, p. 198.

13. Andrassy, op. cit., p. 95. See also Survey of International Law, 1948, UN Doc. A/CN/ 4/1, paragraphs 57-58.

14. Thalmann, op. cit., p. 159 et seq.

15. Yearbook of the International Law Commission, 1949, UN Doc. A/CN. 4/SER. A/1949, p. 288.

16. UN Conference on Water Pollution Problems in Europe, Geneva 1961, Documents submitted to the Conference, vol. II, pp. 450-453, (61 II E/Mim 24).

17. Ibid., vol. III, p. 587.

18. Survey of International Law, 1971, UN Doc. A/CN. 4/245, p. 174.

19. Declaration on the Human Environment, UN Doc. A/CONF. 48/CRP. 26.

20. This paragraph appears also in the preamble to the draft articles mentioned below, in footnotes 50 and 51. Although the Stockholm Declaration in itself is not binding on States, it was the view of several States at the Conference that principle 21 was an affirmation of existing responsibility under International Law.

21. UN Doc. A/RES 2849 (XXVI) of 20 December 1971.

22. For general surveys of the law of marine pollution, see inter alia UN Doc. E/5003 of 7 May 1971, The Sea, and UNITAR Research Report No. 4., "Marine Pollution Problems and Remedies" by Oscar Schachter and Daniel Serwer. See also L'Annuaire de l'Institut de Droit International, 1969, vol. 53-II, pp. 255-343 and 363-369.

23. UNTS, vol. 450, p. 11. Came into force on 3 January 1963. As of 1 April 1971, 48 States were parties.

24. Ibid., vol. 499, p. 311. Came into force on 10 June 1964. As of 1 April 1971, 46 States were parties.

25. UN Doc. A/RES 2467 B (XXIII) of 21 December 1968.

26. UN Doc. A/RES 2566 (XXIV) of 13 December 1969.

27. UN Doc. E/5003 of 7 May 1971, The Sea, paragraph 124.

28. UN Doc. A/RES 2748 (XXV) of 17 December 1970, paragraph 11.

29. UN Doc. A/CONF. 48/IWGMP. II/5 of 22 November 1971. See also A/CONF. 48/8 of January 1972, Identification and Control of Pollutants of Broad International Significance, paragraph 197.

30. See footnote 19; the same text appears in article 11 of the Oslo Convention mentioned in footnote 49.

31. UNTS, vol. 516, p. 205. Came into force on 10 September 1964. As of 1 April 1972, 41 States were parties.

32. Ibid., vol. 559, p. 285. Came into force on 20 March 1966. As of 1 April 1971, 32 States were parties.

33. See above footnote 28.

34. See above, footnote 29.

35. For a discussion on the so called "creeping jurisdiction", see
L. F. E. Goldie, "International Principles of Responsibility for
Pollution", Columbia Journal of Transnational Law, vol. 9, 1970,
pp. 304-306.

36. See above, footnote 24.

37. See above, footnote 29.

38. UN Doc. A/CONF. 48/8 of 7 January 1972, Identification and Control
of Pollutants of Broad International Significance, paragraph 193.
In UN Doc. A/CONF. 48/C. 3/CRP. 30, paragraph 239, the Stockholm Conference recommended that "Governments collectively
endorse the principles set forth in paragraph 197 as guiding concepts
for the Law of the Sea Conference and the IMCO Marine Pollution
Conference scheduled to be held in 1973 and also the statement of
objectives agreed at the second session of the Intergovernmental
Working Group on Marine Pollution as follows:

"The Marine Environment and all the living organisms which it
supports are of vital importance to humanity and all people have
an interest in assuring that this environment is so managed that
its quality and resources are not impaired. This applies especially to coastal nations, which have a particular interest in the management of coastal area resources. The capacity of the sea to assimilate
wastes and render them harmless and its ability to regenerate natural
resources is not unlimited. Proper management is required and
measures to prevent and control marine pollution must be regarded
as an essential element in this management of the oceans and seas
and their natural resources. ",

and in respect of the particular interest of coastal States in the
marine environment and recognizing that the resolution of this
question is a matter for consideration at the Law of the Sea Conferences, take note of the principles on the rights of coastal States
discussed but neither endorsed nor rejected at the second session
of the Intergovernmental Working Group on Marine Pollution and
refer these principles to the 1973 Law of the Sea Conference for
such action as may be appropriate;"

39. UNTS vol. 327, p. 3. Entered into force on 26 July 1958.

40. Ibid. , vol. 600, p. 336.

41. See International Legal Materials, vol. 9, p. 1, 1970.

42. See footnote 23.

43. See e. g. Agreement of 9 June 1969 dealing with Pollution of the
North Sea by Oil. International Legal Materials, vol. 9, p. 359,
1970.

44. Ibid. , vol. 9, p. 25, 1970. Not yet in force.

45. Ibid. , vol. 9, p. 45, 1970. Not yet in force.

46. Ibid. , vol. 11, p. 284, 1972. Not yet in force.

47. J-P. Queneudec, "Les incidences de l'affaire du Torrey Canyon
sur le droit de la mer", Annuaire Français de Droit International,
1967, p. 718.

48. J-P. Queneudec, "Le rejet à la mer de déchets radioactifs", Annuaire Français de Droit International, 1965, p. 778.

49. Convention for the Prevention of Marine Pollution by Dumping from Ships and Aircraft of 15 February 1972. International Legal Materials, vol. 11, p. 262, 1972. The Convention enters into force when seven States have deposited their instruments of ratification. By 1 August 1972, two States have ratified the Convention.

50. See Report of the Intergovernmental Meeting on Ocean Dumping, Document IMOD/4 of 15 April 1972. See also UN Doc. A/CONF/ 48/C. 3/CRP. 30 from the Stockholm Conference where it is recommended in paragraph 233 that Governments with the assistance and guidance of appropriate United Nations bodies, in particular GESAMP, - "refer the draft articles and annexes contained in the report of the intergovernmental meetings in Reykjavik, Iceland, in April 1972 and in London in May 1972 to the United Nations Seabed Committee at its session in July/August 1972 for information and comments and to a conference of Governments to be convened by the Government of the United Kingdom in consultation with the Secretary-General of the United Nations before November 1972 for further consideration with a view to opening the proposed convention for signature at a place to be decided by that Conference, preferably before the end of 1972;" This Conference is expected to take place in November 1972.

51. The same text appears in the UN Declaration on the Human Environment, principle 21.

52. For a general discussion on this subject, see J. C. Dobberts, "Water Pollution and International River Law", Annuaire de l'AAA, vol. 35, 1965, pp. 60-99. See also La Lutte contre la Pollution des Eaux Douces, Conseil de l'Europe, 1966, Michel Despax, La Pollution des eaux et ses problèmes juridiques, Paris 1968, p. 163, G. Sauser-Hall, "L'utilisation industrielle des fleuves internationaux", Hague Recueil, vol. 83, 1953 II, pp. 471-582. A. P. Lester, "River Pollution in International Law", American Journal of International Law, vol. 57, 1963, pp. 828-853, C-A. Colliard, "Evolution et Aspects actuels du régime juridique des fleuves internationaux", Hague Recueil, vol. 125, 1968 III pp. 337-442.

53. See UN River Treaties Collection and UN Rivers Report vols. I, II.

54. Annuaire de l'Institut de Droit International 1911, vol. 24, p. 366. See also 1961, II, vol. 49 p. 370 et seq, article 2: "Tout Etat a le droit d'utiliser les eaux qui traversent ou bordent son territoire sous réserve des limitations imposées par le droit international ... Ce droit a pour limite le droit d'utilisation des autres Etats intéressés au même cours d'eau ou bassin hydrographique. "
(Any State, provided it acts within the restrictions imposed by international law, has the right to use the waterways that cross or border its territory; this right is limited by the rights of other States interested in utilizing the same waterway or hydrographic basin).

55. International Law Association, Report of the fifty-second Conference, Helsinki 1966, p. 477 et seq.

56. O'Connell, International Law, 2nd edition London 1970, p. 593.

57. UN Doc. E/ECE/311, p. 3.

58. Council of Europe, Doc. CE/Mat (67) Misc. 9, European Water Charter.

59. Council of Europe, Doc. CM(70) 134 of 27 October 1970.

60. European Yearbook, 1968, pp. 335-341.

61. See Revue Générale de Droit International Public, 1958, vol. 62,
     pp. 79-123. Sentence du Tribunal arbitral Franco-espagnol en
     date du 16 novembre 1957 dans l'affaire de l'utilisation des eaux
     du lac Lanoux.

62. Ibid. p. 102 and Brunson MacChesney, "Judicial Decisions",
     American Journal of International Law, 1959, p. 156.

63. J. L. Brierly, The Law of Nations, 6th ed. (edited by Sir Humphrey
     Waldock), 1963, pp. 231-33.

64. Reports of International Arbitral Awards, vol. III, p. 1905.

65. See above, p. 2, footnote 6.

66. Ibid. , pp. 1965-66.

67. UN Doc. A/CONF 48/8 of 7 January 1972, Identification and Control
     of Pollutants of Broad International Significance, paragraph 34.

68. Air Pollution across National Boundaries, Sweden's case. Study
     for the United Nations Conference on the Human Environment,
     pp. 87-88.

69. See footnote 23 and below.

70. NATO Doc. C-M (71) 91.

71. OECD Doc. C(72)13 of 10 May 1972. Decision of the Council
     concerning a co-operative technical programme to measure the
     long range transport of air pollutants.

71a. Council of Europe Res. 68(4).

72. Oppenheim op. cit. , p. 291.

73. Cf. footnotes 23 and 69.

74. See inter alia:
     a. Convention on Civil Liability for Nuclear Damage of 21 May
     1963 (IAEA, Vienna). International Legal Materials, vol. 1, p. 727.
     b. Convention on Third Party Liability in the Field of Nuclear
     Energy, of 29 July 1960, (ENEA, Paris) American Journal of
     International Law, vol. 55, 1961, p. 1082.
     c. Convention on Liability of Operators of Nuclear Ships of 25 May
     1962 (IMCO, Brussels) Revue Générale de Droit International Pu-
     blic, vol. 66, 1963, p. 894.
     d. Convention relating to Civil Liability in the Field of Maritime
     Carriage of Nuclear Material of 17 December 1971, International
     Legal Materials, vol. 11, p. 277, 1972.

75. UNTS, vol. 298, p. 169 (Art. 37).

76. See above, footnote 49.

77. See footnote 50. In the draft Annex I "High-level radioactive wastes"
     have been put in sharp brackets.

78. UNTS, vol. 480, p. 43.

79. Declaration on the Human Environment, Principle 26.

80. See inter alia:
     a) Treaty on the Non-proliferation of Nuclear Weapons of 1 July
         1968. International Legal Materials, vol. 7, p. 155.

  b) Treaty on the Prohibition of the Emplacement of Nuclear
   Weapons and other Weapons of Mass Destruction on the
   Seabed and the Ocean Floor and in the Subsoil thereof, (see
   UN Doc. A/RES 2660 (XXV) of 7 December 1970).
  c) Treaty for the Prohibition of Nuclear Weapons in Latin
   America, 1967, UNTS vol. 634, p. 281.
  d) Antarctic Treaty of 1 December 1959, UNTS, vol. 402, p. 71.

81. Verdross, op. cit. , p. 294.

82. For a general introduction on noise pollution, see e. g. Development of the Noise Control Programme, Long Term Programme
  in Environmental Pollution Control Programme, WHO 1971.

83. UNTS, vol. 15, p. 295.

84. International Standards and Recommended Practices, Aircraft Noise,
  Annex 16 to the Convention on International Civil Aviation. Adopted
  by the ICAO Council on 2 April 1971.

85. See e. g. Catherine Delsol, La fin du ciel bleu, les supersoniques,
  Paris 1972. See also Kiss and Lambrechts, "Les dommages causés
  au sol par les vols supersoniques", Annuaire Français de Droit
  International, 1970, pp. 769-781.

86. See e. g. The Danish Act No. 235 of 7 June 1972 on the prohibition
  of civilian flight with supersonic speed.

87. The right to fly in the air above the high seas area is expressly
  mentioned in article 12 in the ICAO convention mentioned in footnote 81, and in article 2 in the Convention of the High Seas, footnote 23.

88. In the Danish Act mentioned above in footnote 86, the ban on the
  flight with supersonic speed only applies to areas within Danish
  jurisdiction.

89. See e. g. Queneudec, op. cit. , p. 758: "Le principe de l'interdiction de pollution est certain an droit international". See also
  Stockholm principle (footnote 19) No. 6 : "The discharge of toxic
  substances, or of other substances and the release of heat, in
  such quantities or concentrations as to exceed the capacity of the
  environment to render them harmless, must be halted in order to
  ensure that serious or irreversible damage is not inflicted upon
  ecosystems. The just struggle of the peoples of all countries
  should be supported".

90. On the sources of Law in the field of transnational pollution, see
  Dobbert, op. cit., pp. 65-83, Andrassy, op. cit. , pp. 97-101,
  Lester op. cit. , passim.

91. See above footnote 55.

92. See above footnote 54.

93. See e. g. the resolutions and declarations referred to in footnotes
  19, 21, 25-29, 38 and 59.

94. See above, under Section II.

95. See e. g. Dobbert op. cit. , p. 79, E. J. de Aréchaga, "International
  responsibility", Manual of Public International Law, edited by Max
  Sørensen, 1968, p. 540, and Lester, op. cit. , pp. 833-36.

96. Inter alia, Andrassy, op. cit. , p. 77 ff, Sauser-Hall, op. cit. ,
  p. 554 ff, Thalmann, op. cit. , passim. and Lester op. cit. , p. 847.

97. Briggs, The Law of Nations, 1952, p. 274 maintains, however, that "no general principle of international law prevents a riparian State .... from diverting or polluting its (sc. the river's) waters." See also the Harmon-doctrine referred to in Dobbert, op. cit., p. 72, and in Lester, op. cit., p. 831.

98. On the organisational question, see Hargrove, op. cit., passim. and UN Doc. A/CONF. 48/11 International Organisational Implications of Action Proposals.

99. UN Doc. A/CONF 48/8, Identification and Control of pollutants of Broad International Significance, paragraphs 124 and 138. See also the UN Declaration on the Human Environment (footnote 19) principle 23:

"23. Without prejudice to such general principles as may be agreed upon by the international community, or to the criteria and minimum levels which will have to be determined nationally, it will be essential in all cases to consider the systems of values prevailing in each country, and the extent of the applicability of standards which are valid for the most advanced countries but which may be inappropriate and of unwarranted social cost for the developing countries."

For a general discussion on pollution standards, see D. Serwer, "International Co-operation for Pollution Control". UNITAR Research Paper No. 9, pp. 5-24.

100. See UN Doc. A/CONF. 48/C.3/CRP. 30, paragraph 220.

101. See above footnote 29.

102. UN Doc. A/RES 2849 (XXVI) of 20 December 1971.

103. UN Doc. A/RES 2850 (XXVI) of 20 December 1971.

104. UN Doc. A/CN. 4/245, Survey of International Law, 1971, paragraph 339.

105. See above, footnote 19.

106. For a discussion on the distinction between wrongful and lawful acts, see UN Doc. A/CN. 4/245, Survey of International Law, pp. 84-94.

107. See e.g. Serwer, op. cit., See also UN Documents E/5003 and A/CONF. 48/8 mentioned above in footnotes 27 and 98.

108. For a general discussion on the procedure for settlement of disputes in the field of pollution, see Dobbert, op. cit., p. 91 ff.

109. See the UN Declaration on the Human Environment (footnote 18) principle 22: "States shall co-operate to develop further the international law regarding liability and compensation in respect of damage which is caused by activities within their jurisdiction or control to the environment of areas beyond their jurisdiction."

For a detailed description of State responsibility for pollution, see Goldie, op. cit., passim. "The Trail Smelter, the Corfu Channel and the Lac Lanoux clearly point to the emergence of strict liability as a principle of public international law"., Ibid., p. 306. See also C.W. Jenks, "Liability for ultra-hazardous activities in international law," Hague Recueil, vol. 117, 1966, pp. 99-200, and Survey of International Law, 1971, UN Doc. A/CN. 4/245, chapter IV.

110. See furthermore John M. Kelson, "State Responsibility and the Abnormally Dangerous Activity", Harvard International Law Journal, Vol. 13, Number 2, Spring 1972, p. 242-43, where the author summarizes the legal situation in the following terms:

"(1) Where the risk of harm from an activity is substantial in either probability or magnitude of harm, and is transnational in character, the State within whose jurisdiction the activity is conducted is under a duty to prevent such harm as may be caused by the enterprise.

(2) A State is under a duty to notify any other State which may be threatened by harm from the abnormally dangerous activities which the State permits to be conducted within its jurisdiction.

(3) A State, failing to prevent harm, shall be originally responsible and strictly liable for the harm caused by abnormally dangerous activities within its jurisdiction to the residents or property of another State."

LEGAL AND INSTITUTIONAL ASPECTS
OF TRANSFRONTIER POLLUTION CONTROL

by

Robert E. Stein

Woodrow Wilson International Center for Scholars,
Washington, D.C., U.S.A.

INTRODUCTION

Accepting the fact of transfrontier pollution, and given the variety
of economic procedures, sanctions, and constraints which may be applied to the problem of transfrontier pollution, it appears only prudent
to consider some of the legal and institutional alternatives as well. The
purpose of this paper is to review the problems and to outline some of
the ways in which members of the international community may wish to
approach transfrontier pollution. Some of the discussion below describes
what is in fact going on, and what international organisations are already
doing to deal with problems in their particular area. However, a great
deal more needs to be done, and States will increasingly be called upon
to exercise their activities not only with a view towards their own well-
being, but with a sense of responsibility for the common good.

STATEMENT OF PROBLEM

Transfrontier pollution may be of several types. It may be considered as wastes being carried to a basin or air shed shared by two or more
countries.[1] It may either be pollution brought from one country to another by air or water, or the common pollution of a river, lake or air
shed which is shared by several countries.[2] In either of these cases,
what is at interest is the ultimate resource. There is little doubt that
in Western Europe as in North America much of the air over major
urban and industrial centers is foul, and large water courses, once
considered the glories of their countries, stink. These sensual signs
are important if only because they graphically demonstrate to a greater
number of people what many know already: that the health and welfare
of their societies is increasingly threatened by the products of the society itself. In many of these areas, it is clear that pollution will not be
totally eliminated. Therefore, in cases of transfrontier pollution, the
countries concerned need first to determine what are the minimal acceptable and optimal levels of pollution.[3] While minimum levels according to health and safety limitations can be drawn globally for a variety of pollutants of the water and air, optimum and attainable levels are
very often better drawn for a smaller region, to take into account more
specific needs.

1. Scott, The Economics of International Transmission of Pollution in "Problems of Environmental Economics", OECD, 1972.
2. Muraro, The Economics of Unidirectional Transfrontier Pollution, this book, p. 33.
3. Smets, Alternative Economic Policies of Unidirectional Transfrontier Pollution, this book,
p. 75.

The great majority of pollution problems that exist today are domestic. A river basin or an air shed may lie entirely within the geographic limits of a specific country. The management and regulation of these problems, therefore, can be also domestic. For other problems of a very similar nature, the river basin or air shed may transcend borders. For these problems of transfrontier pollution domestic solutions may be both ineffective and inadequate. Having identified a pollution problem as being transfrontier in nature, and not completely susceptible to a solution by an individual state, the next question would appear to be how to study, regulate, or manage both the problem and the resource in which the problem exists.[1] It is the best view that States should look to the physical limits of the problem, and plan their pollution control strategy around those limits.[2] As stated by Muraro, the right criterion seems to choose the smallest of the institutions that contains all the countries concerned with the transfrontier pollution.[3] This fairly obvious solution has not, unfortunately, been followed. Recognizing that the Central Commission for the Navigation of the Rhine was not dealing effectively with a variety of land based as well as water based pollution problems of that river, the riparian states in 1963 established a new commission, the International Commission for the Protection of the Rhine Against Pollution. The authority of this Commission, although limited, still only extends to the Rhine River itself rather than to the river and its tributaries. The Commission of the European Communities, in a proposed recommendation from the Council to the Member States of 21st March 1972, stated the necessity to create an European Agency for the Rhine Basin which would be established by public law and would be charged with placing in operation a programme to clean up the waters of the Rhine with the new international Commission itself establishing the programme. This is a further recognition of the principle that the environmental area as a whole should be included in the management of the problem.

Therefore, in considering legal and organisational principles to govern transfrontier pollution, the first principle should be that the problem area as a whole should be included in the consideration of the problem. Stated differently: the approach to be followed by OECD States in their co-operation to ameliorate transfrontier environmental problems should be to include all states which have a direct interest in including a physical relationship to the problem.

PRINCIPLES OF RESPONSIBILITY

There are two aspects to the problem of responsibility. The first is whether and to what extent a state is responsible for pollution damage originating either within its own territory or from an entity subject to

---

1. For a more detailed discussion of focusing on the resource itself see Scott, Economic Aspects of Transnational Pollution, this book, p. 3.

2. For a more detailed discussion of the principle of inclusion of states with a direct interest, see Stein, The Potential of Regional Organisations in Managing the Human Environment, Environmental Series 202, Woodrow Wilson International Center for Scholars (1972).

3. Op. cit. see footnote 2.

its jurisdiction and causes damage outside its territory to another state or entity. Second, if there is such responsibility, under what principles of liability should compensation or other damages be provided to the victim.

At one point it was considered proper that a state could use its territory or its water or its air for any purpose that it deemed fit irrespective of any injury which might occur to another state. That principle is no longer commonly accepted. Rather, there is a recognition in international agreements concluded among many states and in judicial and arbitral decisions in the writing of publicists, that a state cannot use its territory so as to cause injury to others.[1] The traditional starting point for consideration of this principle, is the Trail Smelter Arbitration in which a tribunal examining claims by United States citizens that a smelter in Canada emitted fumes which sent sulphur dioxide across the border causing damage to trees and crops on the U.S. side of the border. The Tribunal held that

> "under principles of international law, as well as the law of the United States, no state has the right to use or permit the use of its territory in such a manner as to cause injury by fumes in or to the territory of another or the properties of persons therein, when the case is of serious consequence and the injury established."[2]

The tribunal did not conclude that the smelter had to shut down its operations, but relied on principles of equity in the establishment of a regime to govern its continued operation. At the same time, the injury was clearly recognized and the regime which was established was designed so as to avoid a continuation of that injury. The extension of this principle is that each state sharing a common resource, be it a water or air shed is entitled to a reasonable and equitable use of that resource. If a state, or entity within a state, engages in activities which deny that reasonable use to another, the Trail Smelter Arbitration can be used as guidance to indicate the responsibility of the offending state for damages. The most recent articulation of that principle was as one of the paragraphs of the Declaration on the Human Environment agreed to by the United Nations Conference on the Human Environment in Stockholm in June of 1972. Paragraph 21 of the Declaration stated in relevant part that

> "States have, in accordance with the Charter of the United Nations and the principles of international law, ... the responsibility to ensure that activities within their jurisdiction or control do not cause damage to the environment of other states or of areas beyond the limits of national jurisdiction."[3]

It was the view of several states at Stockholm, that this principle, rather than breaking new legal ground, was an affirmation of existing responsibility under international law. Therefore, even though the Stockholm Declaration remains to be approved by the UN General Assembly, and even then will not be binding on Member states by force of that approval, this principle as a reflection of existing norms of international law and conduct would be binding on states.

---

1. See Bramsen, Transnational Pollution and International Law and authorities cited therein, this book, p. 257.
2. 3 UN Rep. Int'l. Arb. Awards 1965 (1941).
3. UN Doc. A/CONF. 48/14, p. 7.

This leads to the second major aspect of the question of responsibility: if responsibility does exist, what principles of liability and compensation should be applied with respect to the damage caused. The Stockholm Declaration on the Human Environment also contained a principle dealing with this question. Paragraph 22 states that

"States shall co-operate to develop further the international law regarding liability and compensation for the victims of pollution and other environmental damage caused by activities within the jurisdiction or control of such states to areas beyond their jurisdiction."

What this principle leaves open is the specific type of liability and compensation to be applied. In most circumstances the very fact of pollution which is detectable and identifiable, should be sufficient to sustain liability. This argument for strict liability or liability without fault is receiving increasing favour in discussions of pollution damage. Agreements may either codify or change the principle of liability and may also contain provisions for punitive damages in the event of demonstrated negligence or failure to follow prescribed procedures.

When states do co-operate to develop further the international law regarding liability and compensation, should the principle expressed above that those states directly interested in a specific environmental problem govern the number of states who meet to discuss and agree on the principles of liability and compensation to apply to the governance of a particular problem? That question is in part answered by the following principle of the Stockholm Declaration, paragraph number 23 which states that

"Without prejudice to such general principles as may be agreed upon by the international community, or to the criteria and minimum levels which will have to be determined nationally, it will be essential in all cases to consider the system of values prevailing in each country, and the extent of the applicability of standards which are valid for the most advanced countries but which may be inappropriate and of unwarranted social cost for the developing countries."

The UN Action Paper dealing with pollutants of international significance [1] contained a number of general principles for assessment and control of marine pollution. These principles which had their origin at the meeting of the Second Intergovernmental Working Group in Ottawa, were referred for further consideration to the 1973 IMCO Conference on Marine Pollution and at the Law of the Sea Conference. Ottawa Principle 9 stated that states should join together regionally to concert their policies and adopt measures in common to prevent the pollution of areas which, for geographical or ecological reasons, form a natural entity and an integrated whole. Principle 9 noted that internationally agreed criteria and standards should provide for regional and local variations in the effects of pollution and in the evaluation of these effects. Therefore, it might also make sense to have the states concerned agree to principles of liability and degree of compensation as part of any agreement governing their shared environmental resource.

---

1. UN Doc. A/Conf. 48/8.

There is another aspect of responsibility which was considered at the Stockholm Conference, which has its basis in the OECD's notification and consultation procedure. [1]   The draft Declaration put before the Conference [2]   contained a paragraph which stated

"Relevant information must be supplied by states on activities or developments within their jurisdiction or under their control whenever they believe, or have reason to believe, that such information is needed to avoid the risk of significant adverse effects on the environment in areas beyond their national jurisdiction. "

Disagreement between two Latin American States caused this principle to be referred to the General Assembly for final disposition. It should, however, form a guiding principle for action of states with respect to their responsibility for notifying other states of activities which may have a harmful environmental effect, and which would thus give those other states an opportunity to consult with the acting state on the specific activity. Since the majority of the effects which are international in scope will be regional rather than global, an organisation such as the OECD could expand its existing notification and consultation procedure to include other environmental activities. In addition to being used in this broader scope, the procedures developed by the OECD could be used by other organisations outside of the OECD region. Moreover, an authority or body created by a group of states within the OECD region could as a group agree to utilize the OECD procedures and the organisation itself as a forum for consultation.

Thus far, the situation has been viewed from the perspective of the acting state. But, what of the victim? It is obvious that in many circumstances the acting state may not live up to its responsibility or even if it does, an accident may cause pollution above the agreed to limits. We have already stated that claims for compensation exist, but that may not be sufficient. In the Torrey Canyon accident as in a great number of accidents involving the unintended discharge of pollutants into either the water or the air, some action may be necessary on the part of the possible victim state. As a result of the Torrey Canyon disaster, the IMCO states concluded a Convention in 1969 which gave the coastal state the right to unilaterally take measures to stop pollution of the seas which threaten its shores. [3]   Might not the same principle be applied to inland waters such as major river basins or lakes? It is true, however, that certain examples of air pollution must either be stopped at the source or suffered. For certain cases of air pollution there are no remedies similar to that of an oil spill: air pollution can be stopped at the source, but beyond that usually cannot be contained. For those activities susceptible of remedy after the fact of the accident, both the victim state and other states should formulate joint contingency plans which are discussed in a later section. The right of states to protect their vital interests, sometimes called "self help" must remain circumscribed in that a state's action must be in response to a specific threat, and must be proportional to it. Moreover, one of the primary purposes of concluding the kinds of

1.   "OECD Notification and Consultation Procedure on Measures for Control of Substances Affecting Man or his Environment", C(71)72.

2.   UN Doc. A/Conf. 48/4.

3.   International Convention relating to Intervention on the High Seas in Cases of Oil Pollution Casualties (1969) (Not yet in force).

arrangements discussed in this paper is to enable a state to move from "self help" to an agreed upon response to a previously defined pollution threat.

There are several principles which can be derived from this section. First, states are no longer free to carry out their activities in such a way as to cause injury to other states, for if they do they will be held responsible and required to pay compensation for the damage caused. Second, at the present time the levels of liability or the grounds for compensation remain to be worked out among states concerned with specific problems. Because of the hazardous nature of much of the activity which causes pollution and the resulting damage, liability without fault would appear the most appropriate principle to follow. Third, states should notify other states which might be affected if they are planning to carry out a project or activity which might have a harmful environmental impact on that other state. If a state does notify and solicit consultation from other states with respect to a particular activity and the other state does not object, the question remains whether the acting state is therefore absolved from any responsibility for damage caused to the non-objecting state as a result of the activity. Finally, states should be permitted to take unilateral action to protect their territory from harmful pollution, if no other means to do so have been agreed to.

THE ROLE OF INTERNATIONAL ARRANGEMENTS

The previous sections have outlined the evolving principles of responsibility which will lead towards an increasing number of co-operative efforts to prevent and repair the damage to commonly shared waterways and air spaces. Until now, the majority of organisations which have been given some responsibility for dealing with transfrontier pollution have been largely advisory in nature. Moreover, they have been river or lake agencies, assigned the single task of dealing with the pollution of the water rather than the more general problems of both water and air pollution of the problem area. Additionally, many of these organisations have only recently added pollution control functions to earlier assigned development tasks which have been their primary mission. Thus, for example, the Commissions established in the 19th century to foster navigation of the Rhine and other rivers in Europe have been slow to adapt to the newer problems dealing with pollution. For this reason, on the Rhine as well as in other areas, new commissions have been established with the control of pollution on the river as primary objective. As was noted earlier, a problem with either Rhine Commission, is that they only have responsibility for the river itself rather than a basin approach covering the river and its tributaries. It is also possible for states to agree to act in a certain way with respect to a shared environmental problem without establishing an organisation, agency or commission. Thus, states could agree to adopt compatible legislation to deal with a problem as was done by the Scandanavian countries with respect to oil spills.

A further way of viewing the problem of coping with air or water pollution of a transfrontier nature would be to use a regional approach as a catalyst to broader action. In February 1972, a group of twelve

European states centered around the North-East Atlantic region concluded an agreement dealing with the dumping of harmful substances into the
North Atlantic. Believing that they had a special responsibility with respect to this region, and also unwilling to wait for the conclusion of a
broader global agreement on the subject, the states met in Oslo to conclude an agreement dealing with their particular area.[1]   A permit system was devised utilizing national means of verification which would
require confidence by one state in the ability of another to recognize
the common interest in its granting of the permits. For that reason,
the dumping of certain substances was absolutely prohibited by the states
and a second group of substances requires the permit on a case by case
basis in accordance with agreed procedures. This convention formed
the basis for discussion of a global convention which will be the subject
of a plenepotentiary meeting in London in the Fall of 1972.[2]

If it is decided that an organisation or institution is needed, a
further preliminary question needs to be answered before discussing the
functions to be carried out. Should the institution be single or multi-
purpose; should it be involved in questions of the water of the river basin
or the lake alone or should it rather include under its jurisdiction a
variety of other problems such as air pollution. It would appear that a
uniform answer to this question would not be advisable since the actual
and potential problems of the particular area will differ from place to
place and no hard and fast rule will apply broadly. If the area is one in
which a variety of transfrontier pollution problems exist, the tradeoffs
inherent in the various types of pollution would warrant their being
considered by the same organisation. If, on the other hand, it is decided
to utilize separate organisations each directed towards a single transfrontier pollution problem, it would make sense for an organisation dealing, for example, with a river basin to maintain a working relationship
with an organisation given responsibility for the air shed which overlaps
the river basin so that the problem of tradeoffs and conflicts which might
result can be avoided. The U.S. Canadian International Joint Commission has asked its technical boards concerned with water pollution to
consider the relevance of air pollution in the area to their work on water
pollution.

The following is a list of some of the functions which could well be
given to an international pollution-control institution. The list is not
exclusive, and it does not consider some of the questions of funding, type
of personnel needed and their recruitment. It also does not mention the
research oriented functions which many organisations engage in.

Planning:

The discussion in some of the papers [3] was heavily weighted toward the use of an institution to ameliorate existing pollution and to work
with already serious problems of transfrontier pollution. In large measure, this pollution arose because of lack of resource-wide planning to
allocate the various uses of the resource so as to optimize the continued
use for the common good. Such planning activities, which could be carried

1.   Convention for the Prevention of Marine Pollution by Dumping from Ships and Aircraft, Feb.15,
1972, 11 Int'l Legal Materials 262 (1972).
2.   For a draft of the global convention see UN Doc. A/Conf. 48/8. Add. 1.
3.   By Muraro, Scott and Smets, this book.

out by an international body, would include providing a forum and procedures for notification to other states which share the resource of
planned projects (such as the OECD procedure); resource-wide consideration of land use, and use of site planning for industries and municipalities. Moreover, the adaptation of a plan for rational and environmentally sound growth of the area would enable resource-wide levels of use to
be set which could then be the subject of bargaining and allocation among
the members. Finally, planning would identify the needs of the area for
the training of individuals to monitor and assess the standards, levels
or taxes agreed to by the body or states concerned to regulate transfrontier pollution. It is futile to adopt standards unless the resource can be
adequately monitored and assessed to assure that the standards are
being complied with.

Standard Setting and Apportionment:

A variety of economic and technical functions could be carried out
by the body which would include the identification of the particular pollutants; the minimum or optimum levels which the states wish to achieve
with respect to emissions of those pollutants; the assignment of the right
to emit certain pollutants as well as cost and cost sharing problems of
accepting the standards or taxes which are agreed to. If the agency established to deal with the resource problems of a particular area is without
competence in some of these areas, recourse could be made to a body
such as the OECD for such assistance. It may also be that the OECD
could serve as a more impartial forum for working out disagreements
among the member States.[1]

Monitoring:

A body might be given the responsibility to actually carry out the
monitoring or to co-ordinate the monitoring performed by the member
states of the emissions in the resource area. If monitoring is done on a
constant basis, as is called for in the U.S.-Canadian Great Lakes Water
Quality Agreement,[2] it can also serve as an "early warning system" in
the event of an accident which if not caught in time could cause serious
damage. It would be important for this purpose for the body to have a
cadre of trained individuals whose primary loyalty would be to the body
and not to the particular country from which they came. Any monitoring
performed either by the body or by the states themselves should carry
with it an immediate reporting requirement so that accidents could be
detected immediately. Moreover states should be required to report
immediately to the international body and to the other states any accidents.

Contingency Plans:

It is vitally important that in the event of an accident, common
machinery or parallel machinery can be called into action in order to
relieve the problem as quickly as possible. The U.S.-Canadian Great
Lakes Water Quality Agreement calls for the maintenance of a "joint
U.S.-Canadian contingency plan for the Great Lakes region". The purpose of the plan, according to the agreement, is "to provide for co-

---

1. See section on adjustment of disputes, below.
2. Signed on April 15, 1972, USTIAS No. 7312.

294

ordinated and integrated response to pollution incidents in the Great
Lakes system by responsible federal, state, provincial and local agencies. The plan supplements the national, provincial, and regional plans
of the parties."[1] The objectives of the plan are to develop appropriate
preparedness measures and effective systems for discovery and reporting, to institute prompt measures to restrict spread of the pollutant, and
to provide adequate equipment to respond to pollution incidents. These
plans can be worked out under the authority of the joint body and could
call for a lead agency within each government member of the body to be
the one to organise the domestic response to the accident.

Regulation:

It will be some time before international commissions are given
full responsibility to regulate and enforce violations or breaches of
standards. For that reason, an early task of the body should be to co-
ordinate the regulations of the various member states on a particular
problem, be it air or water quality, to ensure that they are compatible.
The plans for "harmonization" of legislation of the member countries of
the European Communities illustrates the potential of such an activity.
It is important, however, that although compatible though not necessarily
uniform standards may be required throughout the area, provision should
exist for a state with a particular health or welfare problem to adopt
standards stricter than those of the other members of the commission.

ADJUSTMENT OF ENVIRONMENTAL DISPUTES

It is a basic principle of international behaviour and law that states
should first attempt to settle their differences by peaceful means. It is
not as far fetched as it first may appear to foresee the possibility of
armed conflict between two states erupting from an unresolved problem
of transfrontier pollution. Short of that, however, there is a need for a
range of procedures which can be utilized by states and individuals in
them. The procedures can include direct negotiation between the disput-
ing states, the utilization of the good offices of the international body
which manages the resource in question or of another international organisation, the use of mediation and conciliation procedures, arbitration and
judicial settlement. This broad range is important since there are many
disputes the adjustment of which may more easily be achieved by negotiation (bargaining) than through more formal means such as third party
(international) judicial settlement with formal rules of evidence and procedure, and binding judgments. The importance of this subject makes it
suitable for a somewhat detailed treatment.

Before describing some of the aspects of dispute adjustment which
can be utilized, it should be noted that some existing agreements concerned with controlling transfrontier pollution contain provisions for facilitating the adjustment of disputes among the parties. Moreover, there are a
number of broader agreements, such as the Treaty establishing the European Communities which provides for general adjustment procedures.

1. In Annex 8.

There are two distinct avenues which should be encouraged for assisting with the adjustment of disputes arising because of transfrontier pollution or the threat of that pollution.

First, there should be a wider use of domestic procedures for adjusting transfrontier pollution disputes. This includes providing access of nationals from one state to the proceedings in another state which is considering a problem of transfrontier pollution of interest to the first state and its nationals. Procedures should also be accepted in which money judgments rendered in the courts of one state, after a full and fair trial, will be recognized and enforced in the courts of another state. This would permit the victim of transfrontier pollution to obtain a judgment in his courts which will be enforceable against the polluter in his courts where money for satisfaction of a judgment will more likely be found. An extension of this would permit the enforcement of a judgment for equitable relief (e. g. an injunction) prohibiting an action which could cause environmental harm or forbidding the continuance of that action. The use of these procedures is not to encourage "forum shopping", but rather to give the victim a greater opportunity to have his claim heard. If he selects his own courts and is unsuccessful, that fact will certainly be taken into account by the courts in the polluter state if he should then seek access there.

Second, there should be an enlargement of the procedures available to a state which has a shared environmental problem with another state or states and has entered into an agreement with that state or states to manage the problem area. Thus, the U. S. -Canadian Great Lakes Water Quality agreement contains provision for the International Joint Commission to call witnesses and hold hearings on problems of water quality and to engage in independent verification of data or other information submitted by the parties.[1] Another step which an international body might take would be to engage in independent fact-finding (verification) which would be accepted as valid and could be integrated into domestic enforcement procedure. An enlargement of the OECD Notification and Consultation Procedure on Measures for Control of Substances Affecting Man or his Environment would enable states to use the OECD as a forum for the discussion of their disputes with a view to adjusting them and further receiving the guidance of impartial experts in the various aspects of economic analysis which could assist in the resolution of the dispute.

The more formal aspects of dispute adjustment could also be utilized in basin agreements so that states could agree beforehand that a certain principle of liability - e. g. strict liability (liability without fault) - would apply in the event of an accident which causes the levels of transfrontier pollution to exceed the agreed levels. The threat of resort to such provisions may make a state more willing to reach an early accommodation. Moreover, as in the case of the 1971 IMCO agreement on marine pollution, a fund could be established to assist states and persons therein in paying damages in accordance with the agreed-to formula.[2] This agreement could be enforceable in the courts of either the polluter or the victim, whichever forum is most convenient under the circumstances.

1.   Great Lakes Water Quality Agreement, Art. 6.
2.   International Convention on the Establishment of an International Fund for Compensation for Oil Pollution Damage (1971) not yet in force.

Finally, states should not ignore recourse to third party settlement either through arbitration or judicial settlement. Of the two, arbitration would provide a more flexible procedure for working out the dispute. An agreement to submit a dispute to arbitration (with the arbitrators chosen by both parties from a panel of experts in the field of the specific arbitration) would be an equitable and efficient way of reaching a decision binding on the parties to stop a specific activity, reduce or reallocate levels to a new limit, or to pay for an accident. The OECD could offer its services to maintain a file of individuals available for service as arbitrators, with dossiers on each. Moreover, an important part of the success of the Trail Smelter Arbitration Panel's ability to come to grips with the technical aspects of that transfrontier pollution problem was their use of technical advisers, chosen by each of the parties. The OECD could also be helpful in assisting the arbitrators in the selection of these experts especially in the field of economic analysis.

For the near future, states will likely continue to rely primarily on domestic procedures for adjusting their formal disputes. With increased emphasis on the common welfare aspects of transfrontier pollution, there will be an increasing role to be played by any institution established by the parties, as well as by other bodies whose expertise can assist the parties to adjust their differences at as early a stage in the dispute as possible, thus avoiding a bitter divisive confrontation which will not be directed for the common good.

DRAFT GUIDING PRINCIPLES
CONCERNING TRANSFRONTIER POLLUTION

by [1]

Anthony Scott

Professor of Economics,
University of British Columbia, Canada

and

Christopher Bo Bramsen

Lecturer in International Law, University of Copenhagen,
Denmark

1. While the authors take responsibility for the general structure of this paper, they wish to acknowledge the many helpful suggestions and contributions made by the other participants in the Seminar.

A. INTRODUCTION

1.   The OECD shares the growing global concern for the quality of the
environment, and believes that the conservation and improvement of the
environment for the benefit of mankind is an objective which cannot be
obtained without co-operative action by States. It notes with support the
work of the United Nations Conference on the Human Environment; including the important principles set forth in the United Nations Declaration on the Human Environment; and recalls the adoption by the Council
of the OECD on 26 May 1972 of "Guiding Principles Concerning International Economic Aspects of Environmental Policies".

2.   The OECD recognizes its continuing responsibility and that of its
Member countries to deal effectively with the specific problems of pollution, including pollution which transcends national borders, and believes therefore in the need to develop specific procedures, policies and
institutions to ensure efficient and equitable co-operative action to control
transfrontier pollution. For these reasons the OECD recommends that
the Governments of Member countries should, in determining their
environmental control policies and measures with respect to problems
of transfrontier pollution, observe these "Guiding Principles concerning
transfrontier pollution. "

3.   Activities carried out within the national jurisdiction of a State (on
its territory or on ships, oil drilling stations and aircraft registered in
the State) might result in pollution of the environment in areas outside
its jurisdiction (in a foreign territory or in common areas outside any
national jurisdiction, such as the high seas and the air space above).
The pollution that occurs in these cases can be described as transnational,
because it affects areas outside the State's national jurisdiction. Although
a number of common rules can be applied to these different categories
of pollution, the principles described below deal primarily with the case
where pollution originating in one country causes damage in the territory
of another, including territorial waters, after crossing one or more
national frontiers (transfrontier pollution).

4.   For the purpose of these guiding principles, transfrontier pollution can be defined as any discharge of substances or release of energy
originating in the territory of one State that might cause damage, via
natural media (air and water), in the territory of another.

B.   GUIDING PRINCIPLES

5.   It is a rule of international law that States must ensure that activities within their territory do not cause harmful pollution in the territories

of other States and that States are responsible for damage done to the environment of other States. There is, however, no general agreement within the international community on what is meant by harmful pollution. Furthermore, the amount and degree of pollution that is harmful may vary from place to place depending on geographic, social and economic circumstances. It follows that the implementation of this rule of international law calls for joint investigation and action on the part of the States involved.

Co-operation

6.   Co-operation between polluting and polluted States should include consultations leading to agreement on the levels of acceptable pollution, the operation and financing of the necessary pollution prevention and control measures, the procedures for enforcement of the agreement, the assessment and compensation, if any, of damages incurred and the settlement of disputes.

Agreements should also be made concerning measures to prevent, moderate and compensate transfrontier pollution resulting from accidents.

Levels of Pollution

7.   In accordance with the principles of good neighbourliness, States concerned with transfrontier pollution should agree to define levels of pollution acceptable for each pollutant that might cause damage in neighbouring States, taking into account the combined effects of such pollutants. In negotiating tolerable and acceptable levels of pollution, the States should set as their long-term goal the avoidance of environmental degradation creating hazards to human health, harming living resources or damaging social and economic activities, and should take into consideration social, economic and environmental costs of pollution prevention and control.

In establishing levels for transfrontier pollutants that might eventually cause harm to countries not parties to the agreement or to common areas outside national jurisdiction, States should take into account the relevant standards proposed by international organisations. In a situation of unidirectional transfrontier pollution, the polluted State should ensure that pollutants in media flowing into a common area beyond national jurisdiction do not cause damage to a greater extent than that accepted within the agreement area (see explanatory map in Annex III).

The agreement should contain provisions for the periodic review and revision of the levels of pollution.

Joint Resource Quality Management

8.   Attention should be concentrated on the management of rivers, lakes, or airsheds for the joint benefit of the States concerned. Negotiations looking to agreement on the reduction of levels of various pollutants, to efficiency in the extent, location and timing of pollution prevention and control activities, and to objectivity and persistence in data-gathering

are to be preferred to bargaining chiefly about cost, shares and damages. The agreement may focus mainly on a single receiving medium, or broadly on several media within the frontier area.

Agreements concerned with a shared resource entirely within the jurisdiction of two or more States, such as a common lake, river or airshed may, instead of provisions about agreed levels of each pollutant to be achieved by each country, contain provisions about the apportionment among the countries, of an agreed assimilative capacity of the entire resource.

An agency (or agencies) concerned chiefly with the resource and the optimization of its use would be a useful vehicle for channelling national and international concern.

Regional Planning

9.   The agreed levels of pollution depend on the mixtures of industries and activities that exist along the frontier area or along rivers, lakes or in airsheds that cross frontiers. Consequently, the States should consider joint participation in economic and social planning of the entire relevant region, with local participation.

Pollution Prevention and Control Costs

10.   States causing transfrontier pollution should bear all costs for pollution prevention and control measures that are designed to keep the transfrontier pollution within the levels that have been agreed upon by the States concerned. This is an international application of the OECD - "Polluter Pays" - Principle. If a State subjected to transfrontier pollution desires to reduce it to a lower level than that fixed by agreement, it should bear the costs of any additional measures necessary to prevent and control such pollution.

When a shared resource is being polluted by two or more States, agreement should be reached on the apportionment of the total costs of pollution prevention and control.

Compensation for Damages

11.   When damages are caused by transfrontier pollution at a level exceeding the agreed level of pollution, States should be liable to pay compensation for such excess environmental damages to other States.

When damages are caused by transfrontier pollution at a level not exceeding the agreed level of pollution neither the persons carrying out a polluting activity nor the State where such activity is located should be liable to pay compensation for the damages.

When two or more States are responsible for damages caused by transfrontier pollution or have acquired the right to receive compensation for such damages, the compensation paid or received should be divided in an equitable proportion between the States.

When polluting activities have not been the subject of any negotiations, or have been commenced or continued without any prior negotiations having resulted in agreement, the polluting State should, in accordance with international law, pay compensation for all damages caused by the transfrontier pollution.

Economic Efficiency

12. In accordance with Article 2 of the Convention on the OECD in which Members agreed that, while avoiding endangering the economies of other countries, they should individually and jointly promote the efficient use of their economic resources, States concerned with transfrontier pollution should, in approving measures to keep pollution to agreed levels, make serious attempts to use economically efficient methods of pollution prevention and control. In dealing with pollution abatement and damage reduction, States should estimate the respective advantages of locations on both sides of the border, and consider controlling river or lake flows, levels and diversions as well as examine the choice of industrial sites. The strategy adopted need not necessarily involve changing the agreed cost shares, if use is made of financial transfers.

States should consider using discharge fees along with regulations and prohibitions and should undertake joint research on pollution prevention and control measures and mutual adjustment of their activities and anti-pollution measures to transfrontier pollution.

Information

13. Among the essential preconditions for successful agreement on transfrontier pollution is the obtaining and free flow of comprehensive and applicable information and scientific data, accepted by both parties as objective and relevant. To this end, States contemplating negotiation should jointly set up a mechanism such as an expert group which is trusted to advise them on the existing situation and current remedial techniques. Consideration should also be given to the setting up of a centralized data bank by the parties concerned.

Accidents and Contingency Planning

14. States should take all appropriate measures to prevent accidents in their territories that may result in harmful transfrontier pollution and must notify other States of possible sources of such accidents. When activities planned or carried out in a State are likely to cause serious pollution to one or more neighbouring States in case of an accident, the States involved should open consultation with a view to agreeing on the measures to be taken, specifying how damage from such unforeseen and unplanned contingencies should be prevented and contained. Provisions should be made for securing financial compensation and should determine possible maximum amounts to be paid.

15.   In negotiating agreements concerning transfrontier pollution States should include provisions for a range of international dispute adjustment procedures such as mediation, arbitration or judicial settlements.

Such provisions should ensure that nationals of the States concerned be given equal opportunity to seek redress for damage caused by transfrontier pollution. Redress should include compensation and the control of the activity. Where the courts of the polluted State are used, judgements should be enforceable in the polluting State. On a reciprocal basis with a view to incorporating the agreement in the national law, States should consider the adoption of compatible or uniform legislation in pollution matters including legislation governing the right of persons in these cases.

C.   ROLE OF THE OECD

16.   The OECD should provide help and assistance in order to facilitate the implementation of the above guiding principles. Its role would include among others to serve as a forum for the consideration and drafting of a model agreement on transfrontier pollution incorporating the guiding principles described above and containing features described in Annex I. It should make available to Member States its economic expertise and assist them in reaching sound and uniform economic policies which do not cause significant distortions in international trade and investment.

17.   The OECD could also serve as forum for the discussion of a notification and consultation procedure for transfrontier pollution either on an ad hoc basis or pursuant to a general agreement. It could offer its good offices to facilitate the adjustment of disputes involving transfrontier pollution between its Member States and, in this respect, maintain lists and dosiers of economic experts who might be drawn upon by Member States or by arbitrators.

Annex I

CO-OPERATION BETWEEN STATES
ON TRANSFRONTIER POLLUTION PROBLEMS

This annex recommends a procedure for organising international agreement and co-operation among States concerned with a transfrontier pollution problem. The recommendation distinguishes between a long-term programme for resource quality control and arrangements for contingencies and accidents.

A. ORGANISING PROCEDURE

Obtaining sufficient information and achieving an adequate agreement among the States concerned involves the setting-up of two distinct joint international bodies: a Preparatory Commission and its Expert Board.

The Preparatory Commission

The Preparatory Commission is intended as a mechanism for the early stages of diplomatic negotiation. It should ultimately make a joint report to the Governments who have appointed it, and should then expect to be discharged. The Commission's report should constitute a detailed proposal, or draft, of an agreement for both the long-term programme and the contingency arrangements. In order to obtain the technical economic and social information it requires, the Preparatory Commission will set up an Expert Board (described below) and give it terms of reference for its inquiries. The Preparatory Commission may consider a number of different river and airshed transfrontier pollution problems, or be assigned by the countries involved to concentrate on a single resource.

The Preparatory Commission comprises equal numbers of members from each of the countries involved. These members are not expected to be experts in environmental matters, for which knowledge they will depend on the Expert Board. Each member of the Preparatory Commission will be expected both to reflect and defend his country's national interests and to co-operate with other Commission members in drawing up plans which will take into consideration the specific problems and

307

aspirations of the countries involved, the technical possibilities, and the general importance of preventing environmental degradation.

The Expert Board

The Board's function is to produce a report or reports on pollution, damage and abatement as required by the Preparatory Commission. These reports will be based on original research made for the Board and on the Board's collation and processing of existing data. National and other differences on amounts and the benefits and costs of pollution abatement must be reconciled within the Board, who must jointly accept the entire responsibility of adequately informing the Commission. The Board may decide to divide and/or appoint specialist sub-committees or consultants, all reporting to the Board.

The Expert Board is appointed by the Preparatory Commission (which is free to reject names suggested to it by the countries concerned). It includes chemical, civil and sanitary engineers, biological specialists, economists and other professionals, both in the public services and in consulting practice, drawn in equal numbers by the countries involved. The Experts, who may also be drawn from third countries, are chosen in their individual capacities; they should not be advocates for particular points of view.

B.   LONG-TERM RESOURCE QUALITY PROGRAMME

The report of the Preparatory Commission will form a basis for the countries involved to discuss and negotiate the following key elements of a bilateral or multilateral transfrontier resource quality programme:

   i) Agreed levels of pollution: The maximum level of each pollutant that is to be permitted as the medium crosses the frontier from one country into the other.

   ii) Quotas: The levels mentioned above may be expressed as "quotas" in the absorptive or receptive capacity of shared resources or media.

   iii) Timing: The rate or schedule of reduction of each pollutant from existing levels to the final agreed levels: both the length of the transition period and the priority in timing to be given to reduction in the level of each pollutant.

   iv) Agreed abatement measures: Additional to these three elements (levels, quotas and timing) the countries may also agree on specific measures for achieving them, such as the fixing of emission standards for particular processes, locations, or pollutants; the levying of discharge fees; the paying of grants or subsidies; or any combination of these; and may also agree whether particular measures are to be applied in more than one of the countries concerned.

v) Operating agency: The functions, and staffing of an international agency to collect and verify information on the levels, quotas and timing of pollution reduction; the division of responsibility for monitoring between this agency and national bodies; the role of the agency as administrator of water consumption and water level agreements; the membership of the agency and the possibility of converting the Preparatory Commission and/or its Expert Board into an operating agency depending on the functions to be performed; responsiveness to local suggestions and complaints; participation of local authorities in the membership.

vi) Future revision: Provisions for the countries involved to receive recommendations, or negotiate, or both, concerning the revisions of the key elements in their agreement taking account of accumulating information and local experience; and the possible role of the operating agency in expediting this revision.

vii) Cost-Sharing: Agreed shares or formulae for costs of abatement processes and techniques, and of research on these matters; and the procedure for a state to offer, by cost-sharing, to revise the agreed levels, timing or quotas.

viii) Enforcement: Amounts to be paid when agreed pollution levels, schedules or quotas are not achieved; penalties; damages. Other enforcement measures useful to an agency including powers: to hold local hearings; to subpoena witnesses; to question officials; to appoint independent experts for verification; and to issue special public or confidential reports.

ix) Geographical and environmental scope: Coverage of more than one medium or resource, including adjoining land uses, water allocations between countries, and levels and flows of water at the frontier.

x) Individual Rights: The rights of individuals with or without the agency mentioned above, to appear in courts and hearings on either side of the frontier.

Among specific approaches to the long-term resource quality programme, consideration should be given to new mechanisms such as the system of certificates described in the Annex II.

C.   ARRANGEMENTS FOR POLLUTION CONTINGENCIES

Contingencies may occur that raise pollution above the agreed levels for the various pollutants, as a result of waste emissions, oil-spills, pipeline leakages, explosions, fall-out, floods, bad working conditions, use of underqualified personnel, etc.

The report of the Preparatory Commission will form a basis for the countries involved to discuss and negotiate the following key elements of bilateral or multilateral arrangements for contingencies and accidents:

i) <u>Preventive measures</u>: Safety and control measures to prevent accidents resulting from high-risk activities.

ii) <u>Exchange of information</u>: Notification of pollution episodes and accidents, and of dangerous activities taking place within a State, that may lead to transfrontier pollution.

iii) <u>Plans</u>: Procedures for joint action to contain pollution and reduce damage. Emergency rights to enter foreign territory to assist in containing a transfrontier pollution.

iv) <u>Liability</u>: Agreements on liability without fault for super-hazardous activities within the jurisdiction of a State; State responsibility for such activities. Exculpatory circumstances.

v) <u>Compensation</u>: Provisions to ensure the payment of compensation to victims of transfrontier pollution; insurance or other financial security.

vi) <u>Co-ordination</u>: Administration of national and international preventive, notifying, containing and planning activities; damage assessment and compensation by joint operating committees or by a single contingency agency.

vii) <u>Individual rights</u>: The rights of individuals to appear in courts and hearings on either side of the frontier.

POLLUTION CERTIFICATES: A SPECIFIC SYSTEM
OF LONG-TERM RESOURCE-QUALITY CO-OPERATION

States that are unaccustomed to co-operating on international resource management may find it difficult to discuss the essential issues in agreement without acrimony and obstructiveness. In such cases it is helpful to keep in mind the essential elements in negotiation and agreement. The States must know about, and discuss:

i) the amount, timing and nature of present pollution;

ii) the damages (quantitative or at least descriptive);

iii) the cost of abatement of each type of pollution;

iv) the schedule of abatement and which pollutants are to be abated first;

v) the revision and review of the schedules;

vi) cost-sharing.

These elements play a crucial role in the following "certificate" scheme, which can therefore act as a useful centrepiece for international negotiation and bargaining on its details. The scheme also has the merit of providing economic efficiency within the polluting countries in the process of reducing emissions.

The pollution agency would issue "certificates" to each country for the amount of pollution initially discharged. Consequently, the State would be responsible for ensuring that all discharges were discovered, for issuing the certificates to the dischargers, and preventing other uncertified discharges.

Certificates would be issued for specific initial amounts of pollution that may be discharged by each establishment. The amount permitted by a certificate, however, would decline according to a formula to be negotiated. The formula would typically represent a path of maximum discharge from the amount in the initial year to a target amount in a later year. (Examples: a decline of 6% per year for 15 years; a decline of 4 tons per year until zero is reached). The decline rates would be the principal means of cleaning up pollution. They would force a certificate holder either to abate his discharges or to buy extra certificates from another holder, who would be forced to turn to discharge-free processes. The latter feature is described below. Negotiation on the rates of decline would be a principal function of the preparatory commission, after obtaining the necessary initial information.

Certificates would be issued to dischargers of, and for, each type of pollutant. Certificates could be bought and sold at any time, but must be available for sale once a year. Consequently, certificates for different types of pollution (and different locations of discharges) could be exchanged at market prices. This feature would be expected to lead to the highest-cost abatement processes being installed last.

Victims of pollution damage are served in two ways: first, in the negotiated stated rate of decline of permitted discharge; and second, in an additional right to enter the market and buy pollution certificates, thus accelerating the decline in the total amount of pollution permitted. Such revisions in the schedule would cost the victim state whatever the going price in the polluting state.

Another means of empowering intervention by the downstream state is to allow it to issue or sell new pollution rights. In this way, it could sell the right to postpone abatement, if polluters were willing to pay enough.

Purchases of certificates by downstream victim states would provide money to polluters, and would thus be a means of cost-sharing.

Enforcement of pollution reduction would be simplified by the certificate requirement, in that certificate holders would carefully watch one another, looking for evasion.

Annex III
EXPLANATORY MAP

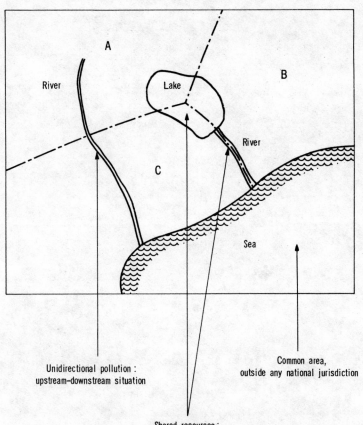

Unidirectional pollution :
upstream–downstream situation

Common area,
outside any national jurisdiction

Shared resources ;
areas entirely within the national jurisdictions
of two or more States

LIST OF PARTICIPANTS

DISCUSSION LEADER

E. Gerelli
Deputy Director
Environment Directorate
OECD, Paris

AUTHORS

d'Arge, Ralph
        Professor, University of California
        Resources for the Future
        1755 Massachusetts Avenue, N.W.
        Washington, D.C.
        U.S.A.

Bramsen, Christopher Bo
        Secretary
        Legal Department
        Danish Ministry of Foreign Affairs
        Copenhagen K.
        Denmark

Cumberland, John H.
        Professor
        University of Maryland
        College Park, Maryland
        U.S.A.

Howlett, Herbert D.
        Chief Engineer
        Delaware River Basin Commission
        P.O. Box 360
        Trenton, New Jersey 08603
        U.S.A.

Kolm, Serge-Christophe
        Professor
        CEPREMAP
        20, rue Henri-Heine
        Paris, 16e
        France

315

Locht, L. J. , Drs.
 Head, Dept of Regional and Project Economics
 Institute for Rural Land and Water Management Research
 Staring gebouw
 Wageningen
 Netherlands

Muraro, Gilberto, Dr.
 Professor
 University of Padua
 Padua
 Italy

Scott, Anthony D.
 Professor
 Department of Economics
 University of British Columbia
 Vancouver 8
 Canada

Smets, Henri, Dr.
 Environment Directorate
 OECD
 2, rue André-Pascal,
 Paris 16e
 France

Stein, Robert
 Fellow
 Woodrow Wilson International Center for Scholars
 Smithsonian Institution Building
 Washington, D.C. 20560
 U.S.A.

OECD SALES AGENTS
DEPOSITAIRES DES PUBLICATIONS DE L'OCDE

ARGENTINA – ARGENTINE
Carlos Hirsch S.R.L.,
Florida 165, BUENOS-AIRES.
☎ 33-1787-2391 Y 30-7122

AUSTRALIA – AUSTRALIE
B.C.N. Agencies Pty, Ltd.,
161 Sturt St., South MELBOURNE, Vic. 3205.
☎ 69.7601
658 Pittwater Road, BROOKVALE NSW 2100.
☎ 938 2267

AUSTRIA – AUTRICHE
Gerold and Co., Graben 31, WIEN 1.
☎ 52.22.35

BELGIUM – BELGIQUE
Librairie des Sciences
Coudenberg 76-78, B 1000 BRUXELLES 1.
☎ 13.37.36/12.05.60

BRAZIL – BRESIL
Mestre Jou S.A., Rua Guaipá 518,
Caixa Postal 24090, 05089 SAO PAULO 10.
☎ 256-2746/262-1609
Rua Senador Dantas 19 s/205-6, RIO DE
JANEIRO GB. ☎ 232-07. 32

CANADA
Information Canada
171 Slater, OTTAWA. KIA 0S9.
☎ (613) 992-9738

DENMARK – DANEMARK
Munksgaards Boghandel
Nørregade 6, 1165 KØBENHAVN K.
☎ (01) 12 69 70

FINLAND – FINLANDE
Akateeminen Kirjakauppa
Keskuskatu 1, 00100 HELSINKI 10. ☎ 625.901

FRANCE
Bureau des Publications de l'OCDE
2 rue André-Pascal, 75775 PARIS CEDEX 16.
☎ 524.81.67
Principaux correspondants :
13602 AIX-EN-PROVENCE : Librairie de
l'Université. ☎ 26.18.08
38000 GRENOBLE : B. Arthaud. ☎ 87.25.11
31000 TOULOUSE : Privat. ☎ 21.09.26

GERMANY – ALLEMAGNE
Verlag Weltarchiv G.m.b.H.
D 2000 HAMBURG 36, Neuer Jungfernstieg 21
☎ 040-35-62-501

GREECE – GRECE
Librairie Kauffmann, 28 rue du Stade,
ATHENES 132. ☎ 322.21.60

ICELAND – ISLANDE
Snaebjörn Jónsson and Co., h.f.,
Hafnarstræti 4 and 9, P.O.B. 1131,
REYKJAVIK. ☎ 13133/14281/11936

INDIA – INDE
Oxford Book and Stationery Co. :
NEW DELHI, Scindia House. ☎ 47388
CALCUTTA, 17 Park Street. ☎ 24083

IRELAND – IRLANDE
Eason and Son, 40 Lower O'Connell Street,
P.O.B. 42, DUBLIN 1. ☎ 01-41161

ISRAEL
Emanuel Brown :
35 Allenby Road, TEL AVIV. ☎ 51049/54082
also at :
9, Shlomzion Hamalka Street, JERUSALEM.
☎ 234807
48 Nahlath Benjamin Street, TEL AVIV.
☎ 53276

ITALY – ITALIE
Libreria Commissionaria Sansoni :
Via Lamarmora 45, 50121 FIRENZE. ☎ 579751
Via Bartolini 29, 20155 MILANO. ☎ 365083
Sous-dépositaires:
Editrice e Libreria Herder,
Piazza Montecitorio 120, 00186 ROMA.
☎ 674628
Libreria Hoepli, Via Hoepli 5, 20121 MILANO.
☎ 865446
Libreria Lattes, Via Garibaldi 3, 10122 TORINO.
☎ 519274
La diffusione delle edizioni OCDE è inoltre assicu-
rata dalle migliori librerie nelle città più importanti.

JAPAN – JAPON
OECD Publications Centre,
Akasaka Park Building,
2-3-4 Akasaka,
Minato-ku
TOKYO 107. ☎ 586-2016
Maruzen Company Ltd.,
6 Tori-Nichome Nihonbashi, TOKYO 103,
P.O.B. 5050, Tokyo International 100-31.
☎ 272-7211

LEBANON – LIBAN
Documènta Scientifica/Redico
Edison Building, Bliss Street,
P.O.Box 5641, BEIRUT. ☎ 354429 – 344425

THE NETHERLANDS – 'PAYS-BAS
W.P. Van Stockum
Buitenhof 36, DEN HAAG. ☎ 070-65.68.08

NEW ZEALAND – NOUVELLE-ZELANDE
The Publications Officer
Government Printing Office
Mulgrave Street (Private Bag)
WELLINGTON, ☎ 46.807
and Government Bookshops at
AUCKLAND (P.O.B. 5344). ☎ 32.919
CHRISTCHURCH (P.O.B. 1721). ☎ 50.331
HAMILTON (P.O.B. 857). ☎ 80.103
DUNEDIN (P.O.B. 1104). ☎ 78.294

NORWAY – NORVEGE
Johan Grundt Tanums Bokhandel,
Karl Johansgate 41/43, OSLO 1. ☎ 02-332980

PAKISTAN
Mirza Book Agency, 65 Shahrah Quaid-E-Azam,
LAHORE 3. ☎ 66839

PHILIPPINES
R.M. Garcia Publishing House,
903 Quezon Blvd. Ext., QUEZON CITY,
P.O. Box 1860 – MANILA. ☎ 99.98.47

PORTUGAL
Livraria Portugal,
Rua do Carmo 70-74. LISBOA 2. ☎ 360582/3

SPAIN – ESPAGNE
Librería Mundi Prensa
Castelló 37, MADRID-1. ☎ 275.46.55
Libreria Bastinos
Pelayo, 52, BARCELONA 1. ☎ 222.06.00

SWEDEN – SUEDE
Fritzes Kungl. Hovbokhandel,
Fredsgatan 2, 11152 STOCKHOLM 16.
☎ 08/23 89 00

SWITZERLAND – SUISSE
Librairie Payot, 6 rue Grenus, 1211 GENEVE 11.
☎ 022-31.89.50

TAIWAN
Books and Scientific Supplies Services, Ltd.
P.O.B. 83, TAIPEI.

TURKEY – TURQUIE
Librairie Hachette,
469 Istiklal Caddesi,
Beyoglu, ISTANBUL, ☎ 44.94.70
et 14 E Ziya Gökalp Caddesi
ANKARA. ☎ 12.10.80

UNITED KINGDOM – ROYAUME-UNI
H.M. Stationery Office, P.O.B. 569, LONDON
SE1 9 NH, ☎ 01-928-6977, Ext. 410
or
49 High Holborn
LONDON WC1V 6HB (personal callers)
Branches at: EDINBURGH, BIRMINGHAM,
BRISTOL, MANCHESTER, CARDIFF,
BELFAST.

UNITED STATES OF AMERICA
OECD Publications Center, Suite 1207,
1750 Pennsylvania Ave, N.W.
WASHINGTON, D.C. 20006. ☎ (202)298-8755

VENEZUELA
Libreria del Este, Avda. F. Miranda 52,
Edificio Galipán, Aptdo. 60 337, CARACAS 106.
☎ 32 23 01/33 26 04/33 24 73

YUGOSLAVIA – YOUGOSLAVIE
Jugoslovenska Knjiga, Terazije 27, P.O.B. 36,
BEOGRAD. ☎ 621-992

Les commandes provenant de pays où l'OCDE n'a pas encore désigné de dépositaire
peuvent être adressées à :
OCDE, Bureau des Publications, 2 rue André-Pascal, 75775 Paris CEDEX 16
Orders and inquiries from countries where sales agents have not yet been appointed may be sent to
OECD, Publications Office, 2 rue André-Pascal, 75775 Paris CEDEX 16

OECD PUBLICATIONS
2, rue André-Pascal, 75775 Paris Cedex 16

———

No. 31.369 1974

PRINTED IN FRANCE